Universitext

Springer Science+Business Media, LLC

Universitext

Editors (North America): S. Axler, F.W. Gehring, and K.A. Ribet

(continued after index)

Chuanming Zong

Sphere Packings

Edited by John Talbot

With 32 Illustrations

Springer

Chuanming Zong
Institute of Mathematics
Chinese Academy of Sciences
Beijing 100080
PR China
cmzong@math08.math.ac.cn

Editor:
John Talbot
Department of Mathematics
University College London
Gower Street
London WC1E 6BT
UK

Editorial Board
(North America):

S. Axler
Mathematics Department
San Francisco State University
San Francisco, CA 94132
USA

F.W. Gehring
Mathematics Department
East Hall
University of Michigan
Ann Arbor, MI 48109-1109
USA

K.A. Ribet
Department of Mathematics
University of California at Berkeley
Berkeley, CA 94720-3840
USA

Mathematics Subject Classification (1991): 05B40, 11H31, 52C17

Library of Congress Cataloging-in-Publication Data
Zong, Chuanming.
    Sphere packings / Chuanming Zong.
        p.   cm. — (Universitext)
        Includes bibliographical references.
        ISBN 978-1-4757-8148-9   ISBN 978-0-387-22780-1 (eBook)
        DOI 10.1007/978-0-387-22780-1
        1. Combinatorial packing and covering. 2. Sphere. I. Title.
    II. Series.
    Qa166.7.Z66   1999
    511'.6—dc21   99-15367

Printed on acid-free paper.

Production managed by MaryAnn Cottone; manufacturing supervised by Jeffrey Taub.
Photocomposed copy prepared from the author's TeX files.

9 8 7 6 5 4 3 2 1

ISBN 978-1-4757-8148-9   SPIN 10715314

To Peter M. Gruber and David G. Larman

Preface

Sphere packings is one of the most fascinating and challenging subjects in mathematics. Almost four centuries ago, Kepler studied the densities of sphere packings and made his famous conjecture. Several decades later, Gregory and Newton discussed the kissing numbers of spheres and proposed the Gregory-Newton problem. Since then, these problems and related ones have attracted the attention of many prominent mathematicians such as Blichfeldt, Dirichlet, Gauss, Hermite, Korkin, Lagrange, Minkowski, Thue, Voronoi, Watson, and Zolotarev, as well as many active today. As work on the classical sphere packing problems has progressed many exciting results have been obtained, ingenious methods have been created, related challenging problems have been proposed and investigated, and surprising connections with other subjects have been found. Thus, though some of its original problems are still open, sphere packings has developed into an important discipline.

There are several books dealing with various aspects of sphere packings. For example, Conway and Sloane [1], Erdös, Gruber, and Hammer [1], L. Fejes Tóth [9], Gruber and Lekkerkerker [1], Leppmeier [1], Martinet [1], Pach and Agarwal [1], Rogers [14], Siegel [2], and Zong [4]. However, none of them gives a full account of this fascinating subject, especially its local aspects, discrete aspects, and proof methods. The purpose of this book is try to do this job. It deals not only with the classical sphere packing problems, but also the contemporary ones such as blocking light rays, the holes in sphere packings, and finite sphere packings. Not only are the main results of the subject presented, but also, its creative methods from areas such as geometry, number theory, and linear programming are described.

In addition, the book contains short biographies of several masters of this discipline and also many open problems.

I am very much indebted to Professors P.M. Gruber, T.C. Hales, M. Henk, E. Hlawka, D.G. Larman, C.A. Rogers, J.M. Wills, G.M. Ziegler, and the reviewers for their helpful information, suggestions, and comments to the manuscript, and to J. Talbot for his wonderful editing work. Nevertheless, I assume sole responsibility for any remaining mistakes. The staff at Springer-Verlag in New York have been courteous, competent, and helpful, especially M. Cottone, K. Fletcher, D. Kramer, and Dr. I. Lindemann. Also, I am very grateful to my wife, Qiaoming, for her understanding and support.

This work is supported by The Royal Society, the Alexander von Humboldt Foundation, and the National Scientific Foundation of China.

Berlin, 1999   C. Zong

Basic Notation

E^n	Euclidean n-dimensional space.
$\mathbf{x}$	A point (or a vector) of E^n with coordinates $(x_1, x_2, \ldots, x_n)$.
$\mathbf{o}$	The origin of E^n.
$\langle \mathbf{x}, \mathbf{y} \rangle$	The inner product of two vectors $\mathbf{x}$ and $\mathbf{y}$.
$\|\mathbf{x}, \mathbf{y}\|$	The Euclidean distance between two points $\mathbf{x}$ and $\mathbf{y}$.
$\|\mathbf{x}\|$	The Euclidean norm of $\mathbf{x}$.
$d(X)$	The diameter of a set X.
$\mathrm{conv}\{X\}$	The convex hull of X.
K	An n-dimensional convex body.
$\mathrm{int}(K)$	The interior of K.
$\mathrm{rint}(K)$	The relative interior of K in the considered space.
$\mathrm{bd}(K)$	The boundary of K.
$v(K)$	The volume of K.
$s(K)$	The surface area of K.
C	An n-dimensional centrally symmetric convex body centered at $\mathbf{o}$.
$\|\mathbf{x}, \mathbf{y}\|_C$	The Minkowski distance between $\mathbf{x}$ and $\mathbf{y}$ with respect to C.
S_n	The n-dimensional unit sphere centered at $\mathbf{o}$.
ω_n	The volume of the n-dimensional unit sphere.
$\|\mathbf{x}, \mathbf{y}\|_g$	The geodesic distance between two points $\mathbf{x}$ and $\mathbf{y}$ in $\mathrm{bd}(S_n)$.
I_n	The n-dimensional unit cube $\{\mathbf{x} : |x_i| \leq \frac{1}{2}\}$.
Z	The set of all integers.

Z_n	The n-dimensional integer lattice $\{\mathbf{z} :\ z_i \in Z\}$.
Λ	An n-dimensional lattice.
$\det(\Lambda)$	The determinant of Λ.
$Q(\mathbf{x})$	A positive definite quadratic form.
γ_n	The Hermite constant.
$\aleph$	A code.
$\|\mathbf{x}, \mathbf{y}\|_H$	The Hamming distance between $\mathbf{x}$ and $\mathbf{y}$.
$D(\mathbf{x})$	The Dirichlet-Voronoi cell at $\mathbf{x}$.
$k(S_n)$	The kissing number of S_n.
$k^*(S_n)$	The lattice kissing number of S_n.
$b(S_n)$	The blocking number of S_n.
$\delta(S_n)$	The maximum packing density of S_n.
$\delta^*(S_n)$	The maximum lattice packing density of S_n.
$\theta(S_n)$	The minimum covering density of S_n.
$\theta^*(S_n)$	The minimum lattice covering density of S_n.
$r(S_n)$	The maximum radius of a sphere that can be embedded into every packing of S_n.
$r^*(S_n)$	The maximum radius of a sphere that can be embedded into every lattice packing of S_n.
$\rho^*(S_n)$	The maximum radius of the spherical base of an infinite cylinder that can be embedded into every lattice packing of S_n.
$h(S_n)$	The Hornich number of S_n.
$h^*(S_n)$	The lattice Hornich number of S_n.
$\ell(S_n)$	The L. Fejes Tóth number of S_n.

Contents

Contents xiii

1. The Gregory–Newton Problem and Kepler's Conjecture

1.1. Introduction

Let K denote a *convex body* in n-dimensional Euclidean space E^n. In other words, K is a compact subset of E^n with nonempty interior such that

$$\lambda\mathbf{x} + (1 - \lambda)\mathbf{y} \in K,$$

whenever both $\mathbf{x}$ and $\mathbf{y}$ belong to K and $0 < \lambda < 1$. We call K a centrally symmetric convex body if there is a point $\mathbf{p}$ such that $\mathbf{x} \in K$ if and only if $2\mathbf{p} - \mathbf{x} \in K$. For example, both the n-dimensional unit sphere

$$S_n = \left\{ \mathbf{x} = (x_1, x_2, \ldots, x_n) : \sum_{i=1}^{n} (x_i)^2 \le 1 \right\}$$

and the n-dimensional unit cube

$$I_n = \left\{ \mathbf{x} = (x_1, x_2, \ldots, x_n) : \max_{1 \le i \le n} |x_i| \le \frac{1}{2} \right\}$$

are centrally symmetric convex bodies. As usual, the interior, boundary, volume, surface area, and diameter of K are denoted by $\mathrm{int}(K)$, $\mathrm{bd}(K)$, $v(K)$, $s(K)$, and $d(K)$, respectively. Denoting the volume of S_n by ω_n, it is well-known that

$$\omega_n = \frac{\pi^{n/2}}{\Gamma(n/2 + 1)},$$

and therefore

$$s(S_n) = n\omega_n = \frac{n\pi^{n/2}}{\Gamma(n/2 + 1)},$$

where $\Gamma(x)$ is the *gamma function*.

If $\mathbf{a}_i = (a_{i1}, a_{i2}, \ldots, a_{in})$, $i = 1, 2, \ldots, n$, are n linearly independent vectors in E^n, then the set

$$\Lambda = \left\{ \sum_{i=1}^{n} z_i \mathbf{a}_i : z_i \in Z \right\},$$

where Z is the set of all integers, is called a *lattice*, and we call $\{\mathbf{a}_1, \mathbf{a}_2, \ldots, \mathbf{a}_n\}$ a *basis* for Λ. As usual, the absolute value of the determinant $|a_{ij}|$ is called the *determinant* of the lattice and is denoted by $\det(\Lambda)$. Geometrically speaking, $\det(\Lambda)$ is the volume of the *fundamental parallelepiped*

$$P = \left\{ \sum_{i=1}^{n} \lambda_i \mathbf{a}_i : 0 \le \lambda_i \le 1 \right\}$$

of Λ.

Let X be a set of discrete points in E^n. We shall call $K + X$ a *translative packing* of K if

$$(\operatorname{int}(K) + \mathbf{x}_1) \cap (\operatorname{int}(K) + \mathbf{x}_2) = \emptyset$$

whenever $\mathbf{x}_1$ and $\mathbf{x}_2$ are distinct points of X. In particular, we shall call it a *lattice packing* of K when X is a lattice.

Let $k(K)$ and $k^*(K)$ denote the *translative kissing number* and the *lattice kissing number* of K, respectively. In other words, $k(K)$ is the maximum number of nonoverlapping translates $K + \mathbf{x}$ that can touch K at its boundary, and $k^*(K)$ is defined similarly, with the restriction that the translates are members of a lattice packing of K. For every convex body K, it is easy to see that

$$k^*(K) \le k(K). \tag{1.1}$$

The following result provides a general upper bound for $k^*(K)$ and $k(K)$.

Theorem 1.1 (Minkowski [7] and Hadwiger [1]). *Let K be an n-dimensional convex body. Then*

$$k^*(K) \le k(K) \le 3^n - 1,$$

where equality holds for parallelepipeds.

Proof. The parallelepiped case is easy to verify. We omit the verification here.

Let $D(A)$ denote the *difference set* of a convex set A. In other words,

$$D(A) = \{\mathbf{x} - \mathbf{y} : \mathbf{x}, \mathbf{y} \in A\}.$$

By convexity, it is easy to see that

$$(A + \mathbf{x}_1) \cap (A + \mathbf{x}_2) \neq \emptyset$$

if and only if

$$\left(\tfrac{1}{2}D(A) + \mathbf{x}_1\right) \cap \left(\tfrac{1}{2}D(A) + \mathbf{x}_2\right) \neq \emptyset.$$

Therefore, by considering the two cases $A = \text{int}(K)$ and $A = K$ it can be easily deduced that

$$k(K) = k(D(K)).$$

Since $D(K)$ is centrally symmetric, in order to get an upper bound for $k(K)$, we assume that K itself is centrally symmetric and centered at $\mathbf{o}$.

If a translate $K + \mathbf{x}$ touches K at $\mathbf{y} \in \text{bd}(K)$, then for any point $\mathbf{z} \in K + \mathbf{x}$, we have

$$3\mathbf{y} \in 3K$$

and

$$\tfrac{3}{2}(\mathbf{z} - \mathbf{y}) \in 3K.$$

Hence we have

$$\mathbf{z} = \tfrac{1}{3} \times 3\mathbf{y} + \tfrac{2}{3} \times \tfrac{3}{2}(\mathbf{z} - \mathbf{y}) \in 3K,$$

and therefore (see Figure 1.1)

$$K + \mathbf{x} \subseteq 3K.$$

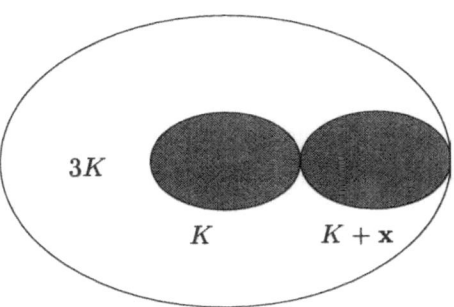

Figure 1.1

Let $K + \mathbf{x}_i$, $i = 1, 2, \ldots, k(K)$, be nonoverlapping translates that touch K at its boundary. Then

$$\bigcup_{i=0}^{k(K)} (K + \mathbf{x}_i) \subseteq 3K,$$

where $\mathbf{x}_0 = \mathbf{o}$, and therefore

$$(k(K) + 1)v(K) \leq v(3K) = 3^n v(K),$$

which implies that

$$k(K) \leq 3^n - 1,$$

and hence using (1.1), Theorem 1.1 is proved.   □

Remark 1.1. *In fact, n-dimensional parallelepipeds are the only convex bodies in E^n such that*

$$k^*(K) = k(K) = 3^n - 1.$$

For a proof see Hadwiger [1] and Groemer [2].

Let l be a positive number and let $m(K, l)$ be the maximum number of translates $K + \mathbf{x}$ that can be packed into the cube lI_n. We define

$$\delta(K) = \limsup_{l \to \infty} \frac{m(K, l)v(K)}{v(lI_n)},$$

the *density* of the *densest translative packings* of K in E^n. Similarly, the density $\delta^*(K)$ of the *densest lattice packings* of K is defined by restricting the translative vectors to those in a lattice. In this case, it can be deduced that

$$\delta^*(K) = \sup_{\Lambda} \frac{v(K)}{\det(\Lambda)},$$

where the supremum is over all lattices Λ such that $K + \Lambda$ is a packing. For every convex body K, it is easy to see that

$$\delta^*(K) \leq \delta(K) \leq 1. \tag{1.2}$$

The density of a given translative packing $K + X$ is defined by

$$\delta(K, X) = \limsup_{l \to \infty} \frac{\text{card}\{X \cap lI_n\}v(K)}{v(lI_n)}.$$

Then simple analysis yields that

$$\delta(K) = \sup_{X} \delta(K, X),$$

where the supremum is over all sets X such that $K + X$ is a packing.

In fact, as the following theorem shows, the density $\delta(K)$ can be defined using any convex body containing the origin, not just the cube.

Theorem 1.2 (Hlawka [4]). *Let A be an arbitrary convex body that contains the origin $\mathbf{o}$ as an interior point, and let $m(K, lA)$ be the maximum number of translates $K + \mathbf{x}$ that can be packed into lA. Then,*

$$\lim_{l \to \infty} \frac{m(K, lA)v(K)}{v(lA)} = \delta(K).$$

Proof. Let $m^*(K, l)$ be the maximum number of nonoverlapping translates $K + \mathbf{x}$ that intersect lI_n and let d be the diameter of K. Since

$$\lim_{l \to \infty} \frac{v((l + 2d)I_n) - v((l - 2d)I_n)}{v(lI_n)} = 0, \tag{1.3}$$

from the definition of $\delta(K)$ it follows that for each $\epsilon > 0$ there is a t_ϵ such that

$$\frac{m(K, t_\epsilon)v(K)}{v(t_\epsilon I_n)} > \delta(K) - \epsilon$$

and

$$\frac{m^*(K, t_\epsilon)v(K)}{v(t_\epsilon I_n)} < \delta(K) + \epsilon.$$

On the other hand, for any convex body A, there exists an $l_\epsilon > 0$ such that for each $l > l_\epsilon$ there are two families of nonoverlapping cubes,

$$\left\{ \tfrac{t_\epsilon}{l} I_n + \mathbf{x}_i \; : \; i = 1, 2, \ldots, g \right\}$$

and

$$\left\{ \tfrac{t_\epsilon}{l} I_n + \mathbf{y}_i \; : \; i = 1, 2, \ldots, h \right\},$$

such that

$$\bigcup_{i=1}^{g} \left(\frac{t_\epsilon}{l} I_n + \mathbf{x}_i \right) \subseteq A,$$

$$v(A) \leq \frac{v \left(\bigcup_{i=1}^{g} \left(\frac{t_\epsilon}{l} I_n + \mathbf{x}_i \right) \right)}{1 - \epsilon},$$

$$A \subseteq \bigcup_{i=1}^{h} \left(\frac{t_\epsilon}{l} I_n + \mathbf{y}_i \right),$$

and

$$v(A) \geq \frac{v \left(\bigcup_{i=1}^{h} \left(\frac{t_\epsilon}{l} I_n + \mathbf{y}_i \right) \right)}{1 + \epsilon}.$$

Hence, when $l > l_\epsilon$, one has

$$(1 - \epsilon)(\delta(K) - \epsilon) < \frac{m(K, lA)v(K)}{v(lA)} < (1 + \epsilon)(\delta(K) + \epsilon),$$

which implies the assertion of Theorem 1.2. □

Remark 1.2. *Clearly, $k(K)$, $k^*(K)$, $\delta(K)$, and $\delta^*(K)$ are invariant under nonsingular linear transformations.*

Suppose that $X = \{\mathbf{x}_1, \mathbf{x}_2, \ldots \}$ is a discrete set of points in E^n. For each point $\mathbf{x}_i \in X$ the *Dirichlet-Voronoi cell* $D(\mathbf{x}_i)$ is defined by

$$D(\mathbf{x}_i) = \{\mathbf{x} : \|\mathbf{x}, \mathbf{x}_i\| \leq \|\mathbf{x}, \mathbf{x}_j\| \text{ for all } \mathbf{x}_j \in X\},$$

where $\|\mathbf{x}, \mathbf{y}\|$ indicates the Euclidean distance between $\mathbf{x}$ and $\mathbf{y}$. In other words,

$$D(\mathbf{x}_i) = \bigcap_{\mathbf{x}_j \in X} \left\{ \mathbf{x} : \langle \mathbf{x} - \mathbf{x}_i, \mathbf{x}_j - \mathbf{x}_i \rangle \leq \tfrac{1}{2} \langle \mathbf{x}_j + \mathbf{x}_i, \mathbf{x}_j - \mathbf{x}_i \rangle \right\},$$

where $\langle \cdot \rangle$ denotes the inner product. Thus every Dirichlet-Voronoi cell is a convex polytope. The Dirichlet-Voronoi cell has many basic properties. For example, it can be easily deduced from its definition that

$$E^n = \bigcup_{\mathbf{x}_i \in X} D(\mathbf{x}_i),$$

and

$$\mathrm{int}(D(\mathbf{x}_j)) \cap \mathrm{int}(D(\mathbf{x}_k)) = \emptyset$$

for every pair of distinct points $\mathbf{x}_j$ and $\mathbf{x}_k$ of X. In other words, the cells $D(\mathbf{x}_i)$ form a *tiling* of E^n. Thus, if $K + X$ is a packing, $D(\mathbf{x}_i)$ can be regarded as a local cell associated to $K + \mathbf{x}_i$. In particular, when K is a sphere centered at $\mathbf{o}$, then

$$K + \mathbf{x}_i \subset D(\mathbf{x}_i)$$

for every point $\mathbf{x}_i \in X$.

By the definition of $\delta(K)$, for any $\epsilon > 0$, there exist a packing $K + X$ and a positive number c such that

$$d(D(\mathbf{x}_i)) < c$$

for every point $\mathbf{x}_i \in X$, and

$$\lim_{l \to \infty} \frac{\mathrm{card}\{X \cap lI_n\} v(K)}{v(lI_n)} \geq \delta(K) - \epsilon.$$

Then, by (1.3), one has

$$\delta(K) \leq \frac{v(K)}{\lambda(X, l)} + 2\epsilon \tag{1.4}$$

for sufficiently large l, where

$$\lambda(X, l) = \frac{\sum_{\mathbf{x}_i \in X \cap lI_n} v(D(\mathbf{x}_i))}{\mathrm{card}\{X \cap lI_n\}}.$$

In particular, since all the Dirichlet-Voronoi cells are congruent if X is a lattice, one has

$$\delta^*(K) = \sup_{\Lambda} \frac{v(K)}{v(D(\mathbf{o}))}, \tag{1.5}$$

where the supremum is over all lattices Λ such that $K + \Lambda$ is a packing. Clearly, (1.4) and (1.5) provide reasonable ways to approximate $\delta(K)$ and $\delta^*(K)$ using local computation.

1.2. Packings of Circular Disks

Lemma 1.1. *Let P be a polygon with p edges containing the unit disk S_2. Then,*

$$v(P) \geq p \tan\frac{\pi}{p},$$

where equality holds if and only if P is regular and its edges are tangent to bd(S_2).

Proof. Without loss of generality, we may assume that the p edges L_1, L_2, ..., L_p of P are tangent to bd(S_2) at t_1, t_2, ..., t_p in a circular order (see Figure 1.2). Letting θ_i be the angle between t_i and t_{i+1}, where $t_{p+1} = t_1$, it follows that

$$\sum_{i=1}^{p} \theta_i = 2\pi$$

and

$$v(P) = \sum_{i=1}^{p} \tan\frac{\theta_i}{2}.$$

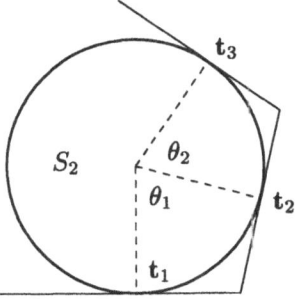

Figure 1.2

Since the function $f(x) = \tan x$ is strictly concave on $[0, \pi/2)$, by *Jensen's inequality* it follows that

$$v(P) \geq p \tan\frac{\sum_{i=1}^{p} \theta_i}{2p} = p \tan\frac{\pi}{p},$$

where equality holds if and only if P is a regular polygon. Lemma 1.1 is proved. $\square$

Let Λ_2 be the two-dimensional lattice with basis $\{(2,0), (1, \sqrt{3})\}$. Then $S_2 + \Lambda_2$ is a lattice packing of S_2. From this example it follows that

$$k^*(S_2) \geq 6 \tag{1.6}$$

and

$$\delta^*(S_2) \geq \frac{v(S_2)}{\det(\Lambda_2)} = \frac{\pi}{\sqrt{12}}. \tag{1.7}$$

On the other hand, if $S_2 + \mathbf{x}_1$ and $S_2 + \mathbf{x}_2$ are two nonoverlapping circular disks that touch S_2 at its boundary, then the angle between $\mathbf{x}_1$ and $\mathbf{x}_2$ is at least $\pi/3$. Therefore,

$$k(S_2) \leq \frac{2\pi}{\pi/3} = 6. \tag{1.8}$$

From (1.1), (1.6), and (1.8) it follows that

$$k^*(S_2) = k(S_2) = 6.$$

The following result determines the exact values of $\delta^*(S_2)$ and $\delta(S_2)$.

Theorem 1.3 (Lagrange [1] and Thue [2]).

$$\delta^*(S_2) = \delta(S_2) = \frac{\pi}{\sqrt{12}}.$$

Proof. Let X be a set such that $S_2 + X$ is a packing and

$$d(D(\mathbf{x})) \leq c$$

for every point $\mathbf{x} \in X$, where c is a suitable positive number. By (1.4)

$$\delta(S_2) \leq \frac{v(S_2)}{\lambda(X, l)} + 2\epsilon \tag{1.9}$$

for sufficiently large l, where

$$\lambda(X, l) = \frac{\sum_{\mathbf{x} \in X \cap lI_2} v(D(\mathbf{x}))}{\text{card}\{X \cap lI_2\}}. \tag{1.10}$$

For convenience, we write

$$X^* = X \cap (l + 2c)I_2$$

and define

$$D^*(\mathbf{x}) = \{\mathbf{y} : \|\mathbf{y}, \mathbf{x}\| \leq \|\mathbf{y}, \mathbf{w}\| \text{ for all } \mathbf{w} \in X^*\}$$

for each $\mathbf{x} \in X^*$. It is easy to see that

$$D^*(\mathbf{x}) = D(\mathbf{x}), \quad \mathbf{x} \in X \cap lI_2, \tag{1.11}$$

and the $D^*(\mathbf{x})$ form a tiling of E^2. Let e, f, and v be the number of edges, polygons, and vertices of this tiling. By *Euler's formula*,

$$f + v = e + 2. \tag{1.12}$$

Let V be the set of vertices of this tiling, and let $p(\mathbf{x})$ and $q(\mathbf{v})$ denote the number of edges of $D^*(\mathbf{x})$ and the number of edges into $\mathbf{v} \in V$, respectively. Since every vertex belongs to at least three edges and every edge joins two vertices, we have

$$3v \le \sum_{\mathbf{v} \in V} q(\mathbf{v}) = 2e. \tag{1.13}$$

Then from (1.12) and (1.13) it follows that

$$e + 6 \le 3f. \tag{1.14}$$

Writing

$$p(X, l) = \frac{\sum_{\mathbf{x} \in X \cap lI_2} p(\mathbf{x})}{\operatorname{card}\{X \cap lI_2\}},$$

(1.3), (1.11), and (1.14) imply that

$$p(X, l) \le \frac{2e}{f} + \epsilon \le 6 - \frac{12}{f} + \epsilon$$

for sufficiently large l, and hence

$$\limsup_{l \to \infty} p(X, l) \le 6. \tag{1.15}$$

By (1.10) and Lemma 1.1 it follows that

$$\lambda(X, l) \ge \frac{\pi \sum_{\mathbf{x} \in X \cap lI_2} \frac{p(\mathbf{x})}{\pi} \tan \frac{\pi}{p(\mathbf{x})}}{\operatorname{card}\{X \cap lI_2\}}.$$

Since the function

$$f(x) = \frac{x}{\pi} \tan \frac{\pi}{x}$$

is concave when $x \ge 3$, by Jensen's inequality and (1.15) it follows that

$$\lambda(X, l) \ge \pi \frac{p(X, l)}{\pi} \tan \frac{\pi}{p(X, l)} \ge \sqrt{12}.$$

Hence by (1.9) one has

$$\delta(S_2) \le \frac{\pi}{\sqrt{12}}. \tag{1.16}$$

Then Theorem 1.3 follows from (1.2), (1.7), and (1.16). $\square$

Remark 1.3. *From the proof of Theorem 1.3 it follows that the lattice with which*

$$\delta^*(S_2) = \frac{\pi}{\sqrt{12}}$$

can be realized is unique with respect to rotation and reflection.

1.3. The Gregory-Newton Problem

Write $\mathbf{a}_1 = (2,0,0)$, $\mathbf{a}_2 = (1,\sqrt{3},0)$, and $\mathbf{a}_3 = (1,\sqrt{3}/3, 2\sqrt{6}/3)$, and let Λ_3 be the lattice generated by them. Usually, Λ_3 is known as the *face-centered cubic lattice*. It is easy to see that $S_3 + \Lambda_3$ is a packing of S_3, in which every sphere touches 12 others. This observation implies

$$k^*(S_3) \geq 12. \tag{1.17}$$

In 1694, during a famous conversation, D. Gregory and I. Newton discussed the following problem.

The Gregory-Newton Problem. *Can a sphere touch 13 spheres of the same size?*

Newton thought "no, the maximum number is 12," while Gregory believed the answer to be "yes." In the literature this problem is sometimes referred to as the *thirteen spheres problem.*

In $S_3 + \Lambda_3$, locally speaking, the kissing configuration is stable. In other words, none of the twelve spheres that touch S_3 can be moved around (see Figure 1.3). However, if twelve unit spheres are placed at positions corresponding to the vertices of a regular *icosahedron* concentric with the central one, the twelve outer spheres do not touch each other and may all be moved around freely (see Figure 1.4).

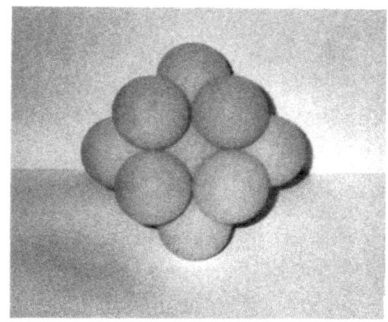 

Figure 1.3   Figure 1.4

Let $S_3 + \mathbf{x}_i$, $i = 1, 2, \ldots, k(S_3)$, be nonoverlapping spheres that touch S_3 at its boundary, and define

$$\Omega_i = \left\{ \mathbf{x} \in \mathrm{bd}(S_3) : \langle \mathbf{x}, \mathbf{x}_i \rangle \geq \sqrt{3} \right\}.$$

The $k(S_3)$ congruent caps Ω_i form a *cap packing* on the surface of S_3, and therefore, comparing the area of bd(S_3) with the area of Ω_i, it follows that

$$k(S_3) \leq \frac{s(S_3)}{s(\Omega_1)} = 14.9282\ldots,$$

which together with (1.17) implies that

$$12 \leq k(S_3) \leq 14.$$

However, this is not enough to solve the Gregory-Newton problem. These facts show the inherent difficulty of this fascinating problem.

Theorem 1.4 (Hoppe [1], Schütte and van der Waerden [1], and Leech [1]).

$$k^*(S_3) = k(S_3) = 12.$$

Proof. Let $X = \{\mathbf{x}_1, \mathbf{x}_2, \ldots, \mathbf{x}_v\}$ be a subset of bd(S_3) such that

$$S_3 + \{\mathbf{o}, 2\mathbf{x}_1, 2\mathbf{x}_2, \ldots, 2\mathbf{x}_v\}$$

is a kissing configuration, and let $\|\mathbf{x}_i, \mathbf{x}_j\|_g$ be the *geodesic distance* between $\mathbf{x}_i$ and $\mathbf{x}_j$. Clearly, we have

$$\|\mathbf{x}_i, \mathbf{x}_j\|_g \geq \pi/3$$

whenever $\mathbf{x}_i$ and $\mathbf{x}_j$ are distinct points of X. Joining $\mathbf{x}_i$ and $\mathbf{x}_j$ by a *geodesic arc* if and only if

$$\|\mathbf{x}_i, \mathbf{x}_j\|_g < \arccos \tfrac{1}{7},$$

we obtain a network on the sphere. No two arcs of this network cross, since any four points of the network form a spherical quadrilateral with side length at least $\pi/3$ whose diagonals cannot both be less than $\pi/2$. Therefore, bd(S_3) is divided into polygons whose vertices are points of X. Clearly, we may assume that every point of X is connected to at least two others by arcs. An easy computation yields that every angle of these polygons is larger than $\pi/3$. Thus at most five arcs of the network meet in any point of X.

Let P_n be a polygon of this network with n sides. By routine computation one can verify the following statements:

$$s(P_3) \geq 0.5512\ldots, \tag{1.18}$$

where equality holds if P_3 is an equilateral triangle with side length $\pi/3$.

$$s(P_4) \geq 1.3338\ldots, \tag{1.19}$$

where equality holds if P_4 is an equilateral quadrilateral with side length $\pi/3$, and one of its diagonals is of length $\arccos \tfrac{1}{7}$.

$$s(P_5) \geq 2.2261\ldots, \tag{1.20}$$

where equality holds if P_5 is an equilateral pentagon with side length $\pi/3$ having two coterminous diagonals of length arccos $\frac{1}{7}$. Also, when $n \geq 4$, it is clear that the excess area of an n-gon over the minimum area of $n-2$ triangles increases with n, since if on a side of any n-gon a triangle of minimum area is abutted, the resulting $(n+1)$-gon has to be deformed so that the original common side becomes a diagonal of length at least arccos $\frac{1}{7}$; consequently, the area is increased.

By (1.12), Euler's formula, one has

$$2v - 4 = 2e - 2f$$
$$= 3f_3 + 4f_4 + 5f_5 + \cdots - 2(f_3 + f_4 + f_5 + \cdots)$$
$$= f_3 + 2f_4 + 3f_5 + \cdots,$$

where e is the number of the arcs, f is the number of the polygons, and f_i is the number of i-gons in the network. Comparing the surface area of S_3 and the area of the network, by (1.18), (1.19), and (1.20) it follows that

$$4\pi \geq 0.5512f_3 + 1.3338f_4 + 2.2261f_5 + \cdots$$
$$= 0.5512(f_3 + 2f_4 + 3f_5 + \cdots) + 0.2314f_4 + 0.5725(f_5 + \cdots)$$
$$= 0.5512(2v - 4) + 0.2314f_4 + 0.5725(f_5 + \cdots).$$

Thus we have

$$2v - 4 \leq \frac{4\pi}{0.5512} = 22.79,$$

and therefore

$$v \leq 13.$$

Suppose now that $v = 13$. Then $2v - 4 = 22$, and we have

$$4\pi \geq 0.5512 \times 22 + 0.2314f_4 + 0.5725(f_5 + \cdots),$$

$$0.44 \geq 0.2314f_4 + 0.5725(f_5 + \cdots),$$

and therefore $f_4 = 0$ or 1 and $f_5 = f_6 = \cdots = 0$. In other words, the network has to divide $\mathrm{bd}(S_3)$ into triangles except possibly for one quadrilateral. Now, we deal with two cases.

Case 1. $f_4 = 0$. Then we have $f_3 = 2e/3$, $13 + 2e/3 = e + 2$, and therefore $e = 33$. Consequently, the average number of arcs meeting at each point $\mathbf{x}_i \in X$ is $66/13 > 5$, which contradicts the fact that at most five arcs meet at any of the v points.

Case 2. $f_4 = 1$. Then it follows from Euler's formula that $f_3 = 20$, $e = 32$, and 4 arcs meet at one point and 5 at every other. In fact, it is impossible to construct such a network. This can be verified by starting from the quadrilateral and adding triangles one by one.

Hence we have proved that

$$k(S_3) \leq 12. \tag{1.21}$$

Then, by (1.1), (1.17), and (1.21) Theorem 1.4 is proved.   □

1.4. Kepler's Conjecture

Routine computation yields that the packing density of $S_3 + \Lambda_3$ is $\pi/\sqrt{18}$, which implies that

$$\delta^*(S_3) \geq \frac{\pi}{\sqrt{18}}. \tag{1.22}$$

Based on this fact and (1.2), J. Kepler in 1611 made the following conjecture.

Kepler's Conjecture.

$$\delta(S_3) = \frac{\pi}{\sqrt{18}}.$$

By (1.4), if the volume of every Dirichlet-Voronoi cell of any packing $S_3 + X$ were greater than or equal to $4\sqrt{2}$, one would be able to deduce

$$\delta(S_3) \leq \frac{\pi}{\sqrt{18}}$$

and hence, by (1.22),

$$\delta^*(S_3) = \delta(S_3) = \frac{\tau}{\sqrt{18}}.$$

Unfortunately, the volumes of some Dirichlet-Voronoi cells of some packings are less than $4\sqrt{2}$. For example, let D be one of the smallest regular *dodecahedra* circumscribed to S_3. Take

$$X = \{o, 2\mathbf{x}_1, 2\mathbf{x}_2, \ldots, 2\mathbf{x}_{12}\},$$

where $\mathbf{x}_i \in \text{bd}(S_3) \cap \text{bd}(D)$. Then, $S_3 + X$ is a packing (see Figure 1.4), and the Dirichlet-Voronoi cell $D(\mathbf{o})$, defined with respect to X, is the dodecahedron D itself. By routine computation, one obtains

$$v(D) = 20 \left(1 - 2\cos\frac{2\pi}{5}\right) \tan\frac{\pi}{5} < 4\sqrt{2}.$$

This example shows the fundamental difficulty of Kepler's conjecture.

Clearly, the density of any sphere packing may be improved by adding spheres as long as there is sufficient room to do so. When there is no longer room to add additional spheres we say that the sphere packing is *saturated.* Without loss of generality, we assume that the sphere packings considered in the following subsections are saturated. We now introduce two approaches to Kepler's conjecture.

1.4.1. L. Fejes Tóth's Program and Hsiang's Approach

Consider the system of Dirichlet-Voronoi cells associated with a sphere packing. If the volume of a Dirichlet-Voronoi cell is less than $4\sqrt{2}$, it seems likely that the volume of some of its neighbors will be larger than $4\sqrt{2}$. Therefore, it is reasonable to consider a locally averaged density, which is what L. Fejes Tóth [9] proposed in 1953 to attack Kepler's conjecture. However, he was unable to realize this program. In 1993, Hsiang [2] announced a proof of Kepler's conjecture. Unfortunately this has been found to contain errors and has not been accepted by experts (see Hales [3]). In fact, the fundamental ideas of both L. Fejes Tóth's program and Hsiang's approach are the same. We only attempt to sketch their key ideas here.

Let h be a fixed number. Two spheres $S_3 + \mathbf{x}_i$ and $S_3 + \mathbf{x}_j$ in a given packing $S_3 + X$ are called *h-neighbors* of each other if $\|\mathbf{x}_i, \mathbf{x}_j\| \le h$. Take

$$X_i = \{\mathbf{x} \in X : \|\mathbf{x}, \mathbf{x}_i\| \le h\}$$

and, for convenience, enumerate it using double indices, namely setting

$$X_i = \{\mathbf{x}_{ij} : j = 1, 2, \ldots, m_i\},$$

where $m_i = \mathrm{card}\{X_i\}$. Then the *h-locally averaged density* of $S_3 + X$ around $S_3 + \mathbf{x}_i$ is defined by

$$\overline{\sigma}(\mathbf{x}_i, X) = \frac{\sum_{j=1}^{m_i} \mu_{ij} \sigma(\mathbf{x}_{ij}, X)}{\sum_{j=1}^{m_i} \mu_{ij}},$$

where

$$\mu_{ij} = \frac{v(D(\mathbf{x}_{ij}))}{m_i}$$

and

$$\sigma(\mathbf{x}_{ij}, X) = \frac{v(S_3)}{v(D(\mathbf{x}_{ij}))}.$$

Lemma 1.2 (Hsiang [2]). *Let $S_3 + X$ be a saturated packing, and set*

$$\overline{\mu}_i = \sum_{j=1}^{m_i} \mu_{ij}$$

to be the weight assigned to $S_3 + \mathbf{x}_i$. Then,

$$\limsup_{l \to \infty} \frac{\sum_{\mathbf{x}_i \in lI_3} \bar{\mu}_i \bar{\sigma}(\mathbf{x}_i, X)}{\sum_{\mathbf{x}_i \in lI_3} \bar{\mu}_i} = \delta(S_3, X).$$

Proof. Since $S_3 + X$ is saturated, the diameter of any Dirichlet-Voronoi cell is bounded by 4 from above. Thus, by (1.4) and the definitions of $\bar{\mu}_i$ and $\bar{\sigma}(\mathbf{x}_i, X)$ it follows that

$$\sum_{\mathbf{x}_i \in lI_3} \bar{\mu}_i \bar{\sigma}(\mathbf{x}_i, X) = \sum_{\mathbf{x}_i \in lI_3} \sum_{j=1}^{m_i} \mu_{ij} \sigma(\mathbf{x}_{ij}, X)$$

$$= \sum_{\mathbf{x}_i \in lI_3} \sum_{j=1}^{m_i} \frac{v(S_3 + \mathbf{x}_{ij})}{m_i}$$

$$= \mathrm{card}\{X \cap lI_3\} v(S_3) + O(l^2)$$

and

$$\sum_{\mathbf{x}_i \in lI_3} \bar{\mu}_i = \sum_{\mathbf{x}_i \in lI_3} \sum_{j=1}^{m_i} \frac{v(D(\mathbf{x}_{ij}))}{m_i}$$

$$= v(lI_3) + O(l^2) .$$

Hence,

$$\limsup_{l \to \infty} \frac{\sum_{\mathbf{x}_i \in lI_3} \bar{\mu}_i \bar{\sigma}(\mathbf{x}_i, X)}{\sum_{\mathbf{x}_i \in lI_3} \bar{\mu}_i} = \limsup_{l \to \infty} \frac{\mathrm{card}\{X \cap lI_3\} v(S_3)}{v(lI_3)}$$

$$= \delta(S_3, X).$$

Lemma 1.2 is proved.   □

From Lemma 1.2 it follows that for any fixed number h,

$$\delta(S_3) \leq \sup_X \sup_{\mathbf{x}_i \in X} \bar{\sigma}(\mathbf{x}_i, X),$$

where the outer supremum is over all sets X such that $S_3 + X$ is a packing. Thus, if it can be proved that for a suitable number h,

$$\bar{\sigma}(\mathbf{x}_i, X) \leq \frac{\pi}{\sqrt{18}} \tag{1.23}$$

holds for every point $\mathbf{x}_i$ of any set X such that $S_3 + X$ is a packing, Kepler's conjecture will follow. In this way, the original global problem has been converted into a local one.

Since $S_3 + X$ is saturated, the number of combinatorial types of the Dirichlet-Voronoi cells is finite. Thus one can try to verify (1.23) by checking a finite number of cases. Clearly, the number of cases depends on the choice of h. In dealing with the individual cases linear programming plays a very important role. L. Fejes Tóth [9] suggested $h = 2.0523\ldots$ in his program. Hsiang [2] took $h = 2.18$ in his approach.

1.4.2. Delone Stars and Hales' Approach

Let X be a set of points such that $S_3 + X$ is a saturated packing. Then E^3 can be decomposed into simplices such that their vertices belong to X and the interior of any sphere circumscribing such a simplex contains no point of X. This decomposition is called the *Delone triangulation*, the simplices are called *Delone simplices*, and the union of the simplices with a common vertex $\mathbf{x}$ is called a *Delone star* (it is denoted by $D^\star(\mathbf{x})$, or simply $D^\star$). To attack Kepler's conjecture by studying Delone stars, Hales [4] took the following approach.

For convenience, we write

$$\beta = \frac{11\pi}{3} - 12\arccos\left(\frac{1}{\sqrt{3}}\right),$$

$$\gamma = \frac{-3\pi + 12\arccos(1/\sqrt{3})}{\sqrt{8}},$$

and

$$F(D^\star) = -v(D^\star)\gamma + \sum_{\mathbf{x}\in X} v\left(D^\star \cap (S_3 + \mathbf{x})\right).$$

Let $D^\star(\mathbf{x})$ be a Delone star with vertices $\mathbf{x}_1, \mathbf{x}_2, \ldots, \mathbf{x}_l$, and denote the intersection of $\mathbf{x}\mathbf{x}_i$ with $\mathrm{bd}(S_3) + \mathbf{x}$ by $\mathbf{p}_i$. Joining $\mathbf{p}_i$ and $\mathbf{p}_j$ by a geodesic arc if and only if

$$\max\left\{\|\mathbf{x}, \mathbf{x}_i\|, \|\mathbf{x}, \mathbf{x}_j\|, \|\mathbf{x}_i, \mathbf{x}_j\|\right\} \leq 2.51,$$

we get a network on $\mathrm{bd}(S_3) + \mathbf{x}$ that divides the surface into a set of regions, say $A_1, A_2, \ldots, A_m$. Let V_i be the cone with vertex $\mathbf{x}$ over A_i, and write

$$C_i^\star = V_i \cap D^\star(\mathbf{x}).$$

The Delone star is divided into m *clusters* $C_1^\star, C_2^\star, \ldots, C_m^\star$. Assign a *score* $\sigma(C_i^\star)$ to every cluster, and define

$$\sigma(D^\star(\mathbf{x})) = \sum_{i=1}^{m} \sigma(C_i^\star)$$

(the exact score for an individual cluster is based on its structure and is very complicated). $\sigma(D^\star(\mathbf{x}))$ is called the score of $D^\star(\mathbf{x})$. It has the following properties.

1. *The score of a cluster depends only on the cluster, and not on the way it sits in a Delone star or in the Delone triangulation of the space.*

2. *If $D^\star$ is a Delone star of the face-centered cubic packing or the hexagonal close packing of S_3, then $\sigma(D^\star) = 8\beta$.*

3. *For any saturated packing $S_3 + X$, we have*

$$\sum_{\mathbf{x} \in X \cap rS_3} \sigma(D^\star(\mathbf{x})) = \sum_{\mathbf{x} \in X \cap rS_3} F(D^\star(\mathbf{x})) + O(r^2). \qquad (1.24)$$

Lemma 1.3 (Hales [4]). *Let $S_3 + X$ be a saturated packing. If*

$$\max_{\mathbf{x} \in X} \{\sigma(D^\star(\mathbf{x}))\} = \eta \le \frac{16\pi}{3},$$

then

$$\delta(S_3, X) \le \frac{16\pi\gamma}{16\pi - 3\eta}.$$

In particular, if $\eta = 8\beta$, then

$$\delta(S_3, X) \le \frac{\pi}{\sqrt{18}}.$$

Proof. Since $S_3 + X$ is a saturated packing,

$$d(D^\star(\mathbf{x})) \le 4$$

holds for every point $\mathbf{x} \in X$. Thus, the number of points $\mathbf{x} \in X$ such that $D^\star(\mathbf{x})$ meets the boundary of rS_3 has order $O(r^2)$. For convenience, we write

$$m(r) = \mathrm{card}\,\{X \cap rS_3\}.$$

Since the Delone stars give a fourfold cover of E^3, by (1.24) we have

$$4\left(-v(rS_3)\gamma + m(r)\frac{4\pi}{3}\right) = \sum_{\mathbf{x} \in X \cap rS_3} \left(-v(D^\star(\mathbf{x}))\gamma\right.$$

$$\left. + \sum_{\mathbf{y} \in X} v(D^\star(\mathbf{x}) \cap (S_3 + \mathbf{y}))\right) + O(r^2)$$

$$= \sum_{\mathbf{x} \in X \cap rS_3} F(D^\star(\mathbf{x})) + O(r^2)$$

$$= \sum_{\mathbf{x} \in X \cap rS_3} \sigma(D^\star(\mathbf{x})) + O(r^2)$$

$$\le m(r)\eta + O(r^2).$$

In other words,

$$m(r)\left(\frac{4\pi}{3} - \frac{\eta}{4}\right) \le v(rS_3)\gamma + O(r^2).$$

Thus,

$$\frac{m(r)v(S_3)}{v(rS_3)} \leq \frac{16\pi\gamma}{16\pi - 3\eta} + O\left(\frac{1}{r}\right).$$

The first part of the lemma follows. Clearly, the second part is the special case $\eta = 8\beta$. Lemma 1.3 is proved. □

In order to use this lemma to prove Kepler's conjecture, Hales [4] proposed a program to verify the following assertions.

Assertion 1.1. *If all the regions are triangles, then* $\sigma(D^\star) \leq 8\beta$.

Assertion 1.2. *If the corresponding region of a cluster* $C^\star$ *has more than three sides, then* $\sigma(C^\star) \leq 0$.

Assertion 1.3. *If all the regions are triangles and quadrilaterals (excluding the case of pentagonal prisms), then* $\sigma(D^\star) \leq 8\beta$.

Assertion 1.4. *If one region has more than four sides, then* $\sigma(D^\star) \leq 8\beta$.

Assertion 1.5. *If* $D^\star$ *is a pentagonal prism, then* $\sigma(D^\star) \leq 8\beta$.

Remark 1.4. *A Delone star is called a pentagonal prism if its regions consist of ten triangles and five quadrilaterals, with the five quadrilaterals in a band around the equator, capped on both ends by five triangles.*

Proofs of these assertions were announced in Hales [4–7] and Ferguson [1], respectively. In fact, Hales [6, 7] and Ferguson [1] employed a different decomposition and a different score system formulated in Ferguson and Hales [1]. However, the basic idea is the same as Hales [4]. Their proofs are extraordinarily complicated. First a computer is used to list all the possibilities for the regions. Then optimization is used to deal with each case. In the course of this proof the computer plays a very important role.

1.5. Some General Remarks

Kepler's conjecture was made in 1611 based on the observation of $S_3 + \Lambda_3$, while the Gregory-Newton problem was proposed in 1694 during a famous discussion. Over the course of the centuries, these problems and their generalizations have attracted the attention of many mathematicians. This section reviews related results that will not be discussed in other chapters.

As with many other problems in geometry, the first significant progress concerning packing densities was made in E^2. In 1773, by studying the minimum of positive binary *quadratic forms* (this method will be discussed in Chapter 2), Lagrange [1] deduced

$$\delta^*(S_2) = \frac{\pi}{\sqrt{12}}.$$ (1.25)

In 1831, Gauss [1] used similar methods to prove a conjecture of Seeber [1] that implies

$$\delta^*(S_3) = \frac{\pi}{\sqrt{18}}.$$ (1.26)

Besides the face-centered cubic lattice packing, for which this density is attained, Barlow [1] found infinitely many nonlattice packings of spheres with the same density. They are the laminations of hexagonal layers of spheres.

Removing the lattice restriction, the first proof of

$$\delta(S_2) = \frac{\pi}{\sqrt{12}}$$ (1.27)

was achieved by Thue [1] and [2]. Roughly speaking, Thue's idea was to compute the area left uncovered by the circular disks in certain finite packings. His method was developed further by Groemer [1] in 1960. Later, different proofs of (1.27) were discovered by L. Fejes Tóth [1], Segre and Mahler [1], Davenport [3], and Hsiang [1]. The proof of Theorem 1.3 is essentially that of L. Fejes Tóth.

Clearly (1.25), (1.26), and (1.27) all support Kepler's conjecture. In 1900, at the International Congress of Mathematicians in Paris, D. Hilbert (see [1]) listed this conjecture as the third part of his 18th problem. Thus it became one of the most popular problems among mathematicians. C.A. Rogers once commented, "Many mathematicians believe, and all physicists know" that the density of the densest sphere packings in E^3 is $\pi/\sqrt{18}$. In 1976, in a review of Hilbert's 18th problem, Milnor [1] wrote on this conjecture, "This is a scandalous situation since the (presumably) correct answer has been known since the time of Gauss (compare Hilbert and Cohn-Vossen). All that is missing is a proof." In 1929, Blichfeldt [1] obtained the first upper bound 0.835 for $\delta(S_3)$. This has been successively improved by Rankin [1], Rogers [10], Lindsey [1], and Muder [1] and [2] to 0.773055.

In 1953, in his well-known book, L. Fejes Tóth suggested a program to attack Kepler's conjecture by checking a finite number of cases. In 1987, Dauenhauer and Zassenhaus [1] made a local approach and suggested the possibility of verifying this conjecture with the help of a computer. In 1993, Hsiang [2] announced a proof using a computer. Unfortunately, it has been found to contain errors (see Hales [3], Hsiang [3], and the preface of Conway and Sloane [1]). Recently, Hales [4–7] and Ferguson [1] announced another proof. To deal with the thousands of cases considered in their proof, even an efficient computer has to run for years.

H. Minkowski was the first mathematician to study the packings of general convex bodies, followed by E. Hlawka, C.A. Rogers, L. Fejes Tóth, and many others. In 1904, Minkowski [5] developed a method by which one can determine the value of $\delta^*(K)$ for any three-dimensional convex body K. As examples he considered the cases of the octahedron and the tetrahedron. Unfortunately, he made a mistake in the tetrahedral case (see Groemer [3]). Later, this mistake was corrected by Hoylman [1]. Although Minkowski's method is important in theory, it was considered not to be practical. However, recently, Betke and Henk [2] used Minkowski's work as a starting point for an algorithm by which one can determine the value of $\delta^*(K)$ for any three-dimensional polytope. As an application, they calculated the value of $\delta^*(K)$ for all regular and Archimedean polytopes.

Around 1950, L. Fejes Tóth [5] and Rogers [4] proved

$$\delta^*(K) = \delta(K)$$

for an arbitrary convex domain in E^2. Thirty years later, L. Fejes Tóth [12] presented an elegant new proof of this result. In high dimensions, it is commonly believed (see Rogers [14], Gruber and Lekkerkerker [1]) that there exist convex bodies K such that

$$\delta^*(K) < \delta(K).$$

Perhaps some high-dimensional spheres have this property. However, no example has been confirmed as yet.

As early as 1896, Minkowski [7] studied lattice packings of a general n-dimensional convex body K, and proved that

$$k^*(K) \leq 3^n - 1,$$

where equality holds if K is a parallelepiped. In 1957 Hadwiger [1] gave an elegant proof (see the proof of Theorem 1.1) of

$$k(K) \leq 3^n - 1,$$

improving the above result. A few years later, Groemer [2] produced a more detailed proof. On the other hand, it was conjectured by Grünbaum [1] that the best lower bound for $k(K)$ is $n(n+1)$, which can be attained by n-dimensional simplices. In 1996, Zong [3] disproved the second part of this conjecture. Very recently, exponential lower bounds for $k(K)$ were discovered by Talata [1] and Larman and Zong [1]. In the plane case, confirming a conjecture of Hadwiger [1], Grünbaum [1] proved that

$$k(K) = k^*(K) = \begin{cases} 8 & \text{if } K \text{ is a parallelogram,} \\ 6 & \text{otherwise.} \end{cases} \qquad (1.28)$$

For a survey on the kissing numbers of general convex bodies we refer to Zong [7].

In the literature R. Hoppe has often been cited as the first mathematician to solve the Gregory-Newton problem. In fact, his proof is not complete. In 1953, using arguments from graph theory, Schütte and van der Waerden [1] completely solved this problem, in favor of Newton. In 1956, Leech [1] gave a new proof. Very recently, Leech's proof was further improved by Aigner and Ziegler [1].

2. Positive Definite Quadratic Forms and Lattice Sphere Packings

2.1. Introduction

There is a remarkable relationship between lattice sphere packings and positive definite quadratic forms. This relationship plays an important role in determining the values of $\delta^*(S_n)$ and $k^*(S_n)$ for small n. Let Λ be a lattice with a basis $\{\mathbf{a}_1, \mathbf{a}_2, \ldots, \mathbf{a}_n\}$, where $\mathbf{a}_i = (a_{i1}, a_{i2}, \ldots, a_{in})$, and write

$$A = \begin{pmatrix} a_{11} & a_{12} & \cdots & a_{1n} \\ a_{21} & a_{22} & \cdots & a_{2n} \\ \vdots & \vdots & \ddots & \vdots \\ a_{n1} & a_{n2} & \cdots & a_{nn} \end{pmatrix}.$$

For convenience, we denote the n-dimensional *integer lattice* by Z_n. There is a positive definite quadratic form related to the lattice Λ,

$$Q(\mathbf{x}) = \langle \mathbf{x}A, \mathbf{x}A \rangle = \mathbf{x}AA'\mathbf{x}',$$

where A' and $\mathbf{x}'$ indicate the transposes of A and $\mathbf{x}$, respectively. Let $r(\Lambda)$ be the largest number such that $r(\Lambda)S_n + \Lambda$ is a packing, let dis(Q) be the *discriminant* of the quadratic form $Q(\mathbf{x})$, and write

$$m(Q) = \min_{\mathbf{z} \in Z_n \setminus \{\mathbf{o}\}} Q(\mathbf{z}).$$

It is easy to see that

$$r(\Lambda) = \tfrac{1}{2}\sqrt{m(Q)}$$

and

$$\text{dis}(Q) = \det(\Lambda)^2.$$

Hence,

$$\delta(r(\Lambda)S_n, \Lambda) = \frac{\omega_n r(\Lambda)^n}{\det(\Lambda)} = \frac{\omega_n m(Q)^{n/2}}{2^n \sqrt{\text{dis}(Q)}}.$$

On the other hand, for each positive definite quadratic form $Q(\mathbf{x}) = \mathbf{x}S\mathbf{x}'$, where S is a positive definite symmetric matrix with entries s_{ij}, there is a corresponding lattice, $\Lambda = \{\mathbf{z}A : \mathbf{z} \in Z_n\}$, such that $S = AA'$. Therefore, there is also a corresponding lattice sphere packing $r(\Lambda)S_n + \Lambda$. Hence, letting $\mathcal{F}$ be the family of all positive definite quadratic forms in n variables, one has

$$\delta^*(S_n) = \sup_{Q(\mathbf{x}) \in \mathcal{F}} \frac{\omega_n m(Q)^{n/2}}{2^n \sqrt{\text{dis}(Q)}}. \tag{2.1}$$

Similarly, letting $M(Q)$ be the set of points $\mathbf{z} \in Z_n \setminus \{\mathbf{o}\}$ at which $Q(\mathbf{z})$ attains its minimum, we have

$$k^*(S_n) = \max_{Q(\mathbf{x}) \in \mathcal{F}} \{\text{card}\{M(Q)\}\}. \tag{2.2}$$

Thus, the geometric problems of determining the values of $\delta^*(S_n)$ and $k^*(S_n)$ are reformulated as arithmetic ones.

Write

$$f(Q) = \frac{m(Q)}{\sqrt[n]{\text{dis}(Q)}},$$

and regard it as a function of the coefficients of $Q(\mathbf{x})$. Then $Q(\mathbf{x})$ is called respectively a *stable form* or an *absolutely stable form* if $f(Q)$ attains a local maximum or an absolute maximum at $Q(\mathbf{x})$. If, in addition, $m(Q) = 1$, then $Q(\mathbf{x})$ is called an *extreme form* or a *critical form*, respectively. Usually, the number

$$\gamma_n = \sup_{Q(\mathbf{x}) \in \mathcal{F}} f(Q) = \sup_{Q(\mathbf{x}) \in \mathcal{F}} \frac{m(Q)}{\sqrt[n]{\text{dis}(Q)}} \tag{2.3}$$

is called *Hermite's constant*. From (2.1) and (2.3) it follows that

$$\delta^*(S_n) = \omega_n \left(\frac{\gamma_n}{4}\right)^{n/2}. \tag{2.4}$$

Moreover, it is easy to see that the density of $r(\Lambda)S_n + \Lambda$ is respectively a local maximum or an absolute maximum if and only if the corresponding form $Q(\mathbf{x})$ is stable or absolutely stable. Therefore, the two problems, of determining the densest lattice sphere packings and determining the critical positive definite quadratic forms, are equivalent.

Let U be a unimodular matrix and write

$$\tilde{Q}(\mathbf{x}) = \mathbf{x}USU'\mathbf{x}'.$$

We say that $\widetilde{Q}(\mathbf{x})$ is equivalent to $Q(\mathbf{x})$. Since the map $\mathbf{z} \mapsto \mathbf{z}U$ is an *automorphism* in Z_n, one has

$$\begin{cases} m(\widetilde{Q}) = m(Q), \\ \text{card}\{M(\widetilde{Q})\} = \text{card}\{M(Q)\}, \\ \text{dis}(\widetilde{Q}) = \det(USU') = \text{dis}(Q). \end{cases} \qquad (2.5)$$

Let $\mathcal{Q}$ be the subfamily of positive definite quadratic forms that are equivalent to $Q(\mathbf{x})$. In other words,

$$\mathcal{Q} = \{\mathbf{x}USU'\mathbf{x}' : U \text{ is a unimodular matrix}\}.$$

Then, the family $\mathcal{F}$ can be represented as a union of different subfamilies $\mathcal{Q}$. So, by (2.5), if in each subfamily $\mathcal{Q}$ a particular form can be chosen, the problem of determining the values of $\delta^*(S_n)$, $k^*(S_n)$, and γ_n using (2.1), (2.2), and (2.3) can be simplified. This is the basic idea of reduction theory. In this chapter the values of $\delta^*(S_n)$, $k^*(S_n)$, and γ_n for small n will be determined using (2.1), (2.2), and (2.3) together with the assistance of different reductions.

2.2. The Lagrange-Seeber-Minkowski Reduction and a Theorem of Gauss

Definition 2.1. *As usual, we denote the greatest common divisor of k integers z_1, z_2, ..., z_k by $(z_1, z_2, \ldots, z_k)$ (here we use bold parentheses). A positive definite quadratic form $Q(\mathbf{x}) = \mathbf{x}S\mathbf{x}'$ is said to be L-S-M reduced if*

$$s_{1j} \geq 0, \quad j = 2, 3, \ldots, n,$$

and

$$Q(\mathbf{z}) \geq s_{ii} \qquad (2.6)$$

for all integer vectors $\mathbf{z} = (z_1, z_2, \ldots, z_n)$ such that

$$(z_i, z_{i+1}, \ldots, z_n) = 1, \quad i = 1, 2, \ldots, n.$$

Lemma 2.1 (Minkowski [6]). *Every positive definite quadratic form is equivalent to an L-S-M reduced one.*

Proof. Let $Q(\mathbf{x}) = \mathbf{x}S\mathbf{x}'$ be a positive definite quadratic form. Using basic algebra we know that there exists a nonsingular matrix A such that $S = AA'$. Denote the lattice $\{\mathbf{z}A : \mathbf{z} \in Z_n\}$ by Λ. Then $Q(\mathbf{z})^{1/2}$ is the Euclidean norm of the vector $\mathbf{z}A$ in Λ. Hence, Lemma 2.1 can be proved by showing that the following assertion holds.

Assertion 2.1. *Every lattice Λ has a basis $\{\mathbf{a}_1, \mathbf{a}_2, \ldots, \mathbf{a}_n\}$ such that*

$$\left\langle \sum_{j=1}^{n} z_j \mathbf{a}_j, \sum_{j=1}^{n} z_j \mathbf{a}_j \right\rangle \geq \langle \mathbf{a}_i, \mathbf{a}_i \rangle \tag{2.7}$$

for all integer vectors $\mathbf{z} = (z_1, z_2, \ldots, z_n)$ with

$$(z_i, z_{i+1}, \ldots, z_n) = 1, \quad i = 1, 2, \ldots, n,$$

and

$$\langle \mathbf{a}_1, \mathbf{a}_j \rangle \geq 0, \quad j = 2, 3, \ldots, n. \tag{2.8}$$

We now prove this assertion. First, we take $\mathbf{a}_1$ to be a point of Λ satisfying

$$\langle \mathbf{a}_1, \mathbf{a}_1 \rangle = \min_{\mathbf{a} \in \Lambda \setminus \{\mathbf{o}\}} \langle \mathbf{a}, \mathbf{a} \rangle.$$

Let B_2 be the set of points $\mathbf{a} \in \Lambda$ such that $\{\mathbf{a}_1, \mathbf{a}\}$ can be extended to a basis for Λ. From linear algebra we know that the set B_2 is nonempty. Then, take $\mathbf{a}_2$ to be a point of B_2 satisfying $\langle \mathbf{a}_1, \mathbf{a}_2 \rangle \geq 0$ and

$$\langle \mathbf{a}_2, \mathbf{a}_2 \rangle = \min_{\mathbf{a} \in B_2} \langle \mathbf{a}, \mathbf{a} \rangle.$$

Inductively, assume that $i < n$ and $\mathbf{a}_1, \mathbf{a}_2, \ldots, \mathbf{a}_i$ have been chosen. Let B_{i+1} be the set of points $\mathbf{a} \in \Lambda$ such that $\{\mathbf{a}_1, \mathbf{a}_2, \ldots, \mathbf{a}_i, \mathbf{a}\}$ can be completed to a basis for Λ. Note that again we know that B_{i+1} is nonempty. Take $\mathbf{a}_{i+1}$ to be a point of B_{i+1} satisfying $\langle \mathbf{a}_1, \mathbf{a}_{i+1} \rangle \geq 0$ and

$$\langle \mathbf{a}_{i+1}, \mathbf{a}_{i+1} \rangle = \min_{\mathbf{a} \in B_{i+1}} \langle \mathbf{a}, \mathbf{a} \rangle. \tag{2.9}$$

It is easy to see that $\{\mathbf{a}_1, \mathbf{a}_2, \ldots, \mathbf{a}_n\}$ is a basis of Λ that satisfies (2.8). Let $\mathbf{a} = \sum_{j=1}^{n} z_j \mathbf{a}_j$ be a point of Λ such that $(z_i, z_{i+1}, \ldots, z_n) = 1$. From linear algebra we know that $\{\mathbf{a}_1, \mathbf{a}_2, \ldots, \mathbf{a}_i, \mathbf{a}\}$ can be extended to a basis for Λ. In other words, $\mathbf{a} \in B_{i+1}$. Then, (2.7) follows from (2.9). Therefore, Assertion 2.1 and Lemma 2.1 are proved.   □

In fact, by a theorem of Minkowski [6], in order to verify that a positive definite quadratic form is L-S-M reduced one only need check that (2.6) holds for a finite number of integer vectors. Here we introduce two practical criteria for the L-S-M reduced binary forms and the L-S-M reduced ternary forms.

Lemma 2.2 (Lagrange [1]). *A positive definite binary quadratic form $Q(\mathbf{x}) = \mathbf{x}S\mathbf{x}'$ is L-S-M reduced if and only if*

$$\begin{cases} s_{11} \leq s_{22}, \\ 0 \leq 2s_{12} \leq s_{11}. \end{cases} \tag{2.10}$$

Proof. Let $Q(\mathbf{x})$ be a form satisfying (2.10). Then, for $\mathbf{z} = (z_1, z_2) \in Z_2$,

$$
\begin{aligned}
Q(\mathbf{z}) &= s_{11}(z_1)^2 + 2s_{12}z_1z_2 + s_{22}(z_2)^2 \\
&\geq s_{11}(z_1)^2 - s_{11}|z_1z_2| + s_{11}(z_2)^2 + (s_{22} - s_{11})(z_2)^2 \\
&= s_{11}\left[(z_1)^2 - |z_1z_2| + (z_2)^2\right] + (s_{22} - s_{11})(z_2)^2.
\end{aligned}
$$

Clearly, for $\mathbf{z} \neq \mathbf{o}$,

$$(z_1)^2 - |z_1z_2| + (z_2)^2 \geq 1.$$

Therefore, since $s_{22} \geq s_{11}$, we have

$$
Q(\mathbf{z}) \geq \begin{cases} s_{11}, & \text{if } \mathbf{z} \neq \mathbf{o}, \\ s_{22}, & \text{if } z_2 \neq 0. \end{cases}
$$

Then, it follows from Definition 2.1 that $Q(\mathbf{x})$ is L-S-M reduced.

On the other hand, if $Q(\mathbf{x})$ is L-S-M reduced, the inequalities of (2.10) follow from $Q(1,0) \leq Q(0,1)$ and $Q(0,1) \leq Q(1,-1)$. Hence, Lemma 2.2 is proved. □

Remark 2.1. *Let $Q(\mathbf{x})$ be an L-S-M reduced positive definite binary quadratic form. By (2.10), we have*

$$
\frac{\text{dis}(Q)}{s_{11}s_{22}} = \frac{s_{11}s_{22} - (s_{12})^2}{s_{11}s_{22}} \geq \frac{s_{11}s_{22} - (s_{11})^2/4}{s_{11}s_{22}}
$$

$$
\geq 1 - \frac{s_{11}}{4s_{22}} \geq \frac{3}{4}.
$$

Hence,

$$
m(Q) = s_{11} \leq \sqrt{s_{11}s_{22}} \leq \sqrt{\tfrac{4}{3}\text{dis}(Q)}.
$$

Then, by (2.1) and certain example it follows that

$$
\delta^*(S_2) = \frac{\pi}{\sqrt{12}}.
$$

Thus we obtain an arithmetic proof for the lattice case of Theorem 1.3. Examining the above argument it is easy to see that every critical positive definite binary quadratic form is equivalent to

$$
Q(\mathbf{x}) = (x_1)^2 + x_1x_2 + (x_2)^2,
$$

which implies the uniqueness of the densest lattice circle packing. This idea goes back to J.L. Lagrange.

Lemma 2.3 (Seeber [1]). *A positive definite ternary quadratic form $Q(\mathbf{x}) = \mathbf{x}S\mathbf{x}'$ is L-S-M reduced if and only if*

$$
\begin{cases}
s_{11} \leq s_{22} \leq s_{33}, \\
0 \leq 2s_{12} \leq s_{11}, \\
0 \leq 2s_{13} \leq s_{11}, \\
0 \leq 2|s_{23}| \leq s_{22}, \\
-2s_{23} \leq s_{11} + s_{22} - 2(s_{12} + s_{13}).
\end{cases} \tag{2.11}
$$

Proof. For convenience, let (2.11.i) indicate the ith inequality of (2.11). Assume that $Q(\mathbf{x})$ satisfies (2.11), we proceed to show that it is L-S-M reduced. In other words,

$$Q(\mathbf{z}) \geq \begin{cases} s_{11}, & \text{if } (z_1, z_2, z_3) = 1, \\ s_{22}, & \text{if } (z_2, z_3) = 1, \\ s_{33}, & \text{if } z_3 \neq 0. \end{cases} \tag{2.12}$$

If $z_3 = 0$, then by (2.11.1) and (2.11.2), $Q(z_1, z_2, 0)$ is L-S-M reduced. Then, the first two inequalities of (2.12) follow from Lemma 2.2. Now we proceed to show the last inequality of (2.12) by dealing with two cases.

Case 1. $z_3 \neq 0$ *and* $z_1 = 0$. Define $\theta = 1$ if $s_{23} \geq 0$, and $\theta = -1$ if $s_{23} < 0$. From (2.11.1) and (2.11.4) it follows that $Q(0, z_2, \theta z_3)$ is L-S-M reduced. Thus, by Lemma 2.2, $Q(0, z_2, z_3) \geq s_{33}$. Similarly, it can be proved that $Q(z_1, 0, z_3) \geq s_{33}$.

Case 2. $z_1 z_2 z_3 \neq 0$. Routine computation yields that

$$\begin{aligned} Q(\mathbf{z}) &= (s_{11} - s_{12} - s_{13})(z_1)^2 + (s_{22} - s_{12} - |s_{23}|)(z_2)^2 \\ &\quad + (s_{33} - s_{13} - |s_{23}|)(z_3)^2 + \Sigma, \end{aligned} \tag{2.13}$$

where

$$\Sigma = s_{12}(z_1 + z_2)^2 + s_{13}(z_1 + z_3)^2 + |s_{23}|(z_2 + \theta z_3)^2.$$

From the first four inequalities of (2.11) it follows that all the coefficients of (2.13) and Σ are nonnegative. Assume that $s_{12} \leq s_{13} \leq |s_{23}|$. We consider three subcases.

i. *At least two of the three terms* $z_1 + z_2$, $z_1 + z_3$, *and* $z_2 + \theta z_3$ *are nonzero.* Then $\Sigma \geq 2s_{12}$, and therefore, by (2.13), (2.11.3), and (2.11.4),

$$\begin{aligned} Q(\mathbf{z}) &\geq s_{11} + s_{22} + s_{33} - 2(s_{12} + s_{13} + |s_{23}|) + 2s_{12} \\ &= (s_{11} - 2s_{13}) + (s_{22} - 2|s_{23}|) + s_{33} \\ &\geq s_{33}. \end{aligned}$$

ii. *Exactly two of the three terms* $z_1 + z_2$, $z_1 + z_3$, *and* $z_2 + \theta z_3$ *are zero.* Then, the third term must be a multiple of two, and therefore $\Sigma \geq 2s_{12}$. Similarly to the previous subcase we have $Q(\mathbf{z}) \geq s_{33}$.

iii. $z_1 + z_2 = z_1 + z_3 = z_2 + \theta z_3 = 0$. Then $\theta = -1$ and $s_{23} < 0$. Thus, by (2.13) and (2.11.5),

$$\begin{aligned} Q(\mathbf{z}) &= s_{11} + s_{22} + s_{33} - 2(s_{12} + s_{13} - s_{23}) \\ &= s_{33} + 2s_{23} + [s_{11} + s_{22} - 2(s_{12} + s_{13})] \\ &\geq s_{33}. \end{aligned}$$

On the other hand, if $Q(\mathbf{x})$ is an L-S-M reduced ternary form, then (2.11) follows from the inequalities $Q(1,0,0) \leq Q(0,1,0) \leq Q(0,0,1)$,

$Q(0,1,0) \leq Q(1,-1,0)$, $Q(0,0,1) \leq Q(1,0,-1)$, $Q(0,0,1) \leq Q(0,1,\pm 1)$, and $Q(0,0,1) \leq Q(-1,1,1)$. Hence, Lemma 2.3 is proved. $\square$

Theorem 2.1 (Gauss [1]).

$$\delta^*(S_3) = \frac{\pi}{\sqrt{18}}. \tag{2.14}$$

Furthermore, the densest lattice packing of S_3 is unique up to rotation and reflection.

Proof. By (2.1), (2.5), Lemma 2.1, and Lemma 2.3, in order to prove (2.14) it is sufficient to show that

$$\text{dis}(Q) \geq \tfrac{1}{2}s_{11}s_{22}s_{33} \tag{2.15}$$

for every positive definite ternary quadratic form $Q(\mathbf{x}) = \mathbf{x}S\mathbf{x}'$ that satisfies (2.11). It is well-known that

$$\text{dis}(Q) = s_{11}s_{22}s_{33} + 2s_{12}s_{13}s_{23} - s_{11}(s_{23})^2 - s_{22}(s_{13})^2 - s_{33}(s_{12})^2. \tag{2.16}$$

Fixing s_{11}, s_{22}, and s_{33}, and regarding

$$\text{dis}(Q) = f(s_{12}, s_{13}, s_{23})$$

as a function of s_{12}, s_{13}, and s_{23}, we proceed to prove (2.15) by dealing with two cases.

Case 1. $s_{23} < 0$. Since $\text{dis}(Q)$ is a convex function of s_{23}, its minimum, under the constraints of (2.11), is attained at

$$s_{23} = \min \left\{ s_{12} + s_{13} - \tfrac{1}{2}(s_{11} + s_{22}), -\tfrac{1}{2}s_{22} \right\}.$$

Thus, this case can be dealt with in two subcases.

i. $s_{23} = s_{12} + s_{13} - (s_{11} + s_{22})/2$. Then, it follows from (2.16) that

$$\text{dis}(Q) = a_1(s_{13})^2 + b_1 s_{13} + c_1, \tag{2.17}$$

where a_1, b_1, and c_1 are independent of s_{13}, and

$$a_1 = -[(s_{11} - 2s_{12}) + s_{22}] < 0.$$

In addition, by (2.11.4) and the assumption of this subcase,

$$s_{11} - 2s_{12} \leq 2s_{13}. \tag{2.18}$$

By (2.11.3) the minimal value of (2.17) is attained either at $s_{13} = 0$ or at $s_{13} = s_{11}/2$. When $s_{13} = 0$, it follows from (2.11.2), (2.18), and the assumption of this subcase that $s_{12} = s_{11}/2$ and $s_{23} = -s_{22}/2$, which will be dealt with in the following subcase. When $s_{13} = s_{11}/2$ we have

$$\mathrm{dis}(Q) = a_2(s_{12})^2 + b_2 s_{12} + c_2, \tag{2.19}$$

where a_2, b_2, and c_2 are independent of s_{12}, and $a_2 = -s_{33} < 0$. By (2.11.2) the minimal value of (2.19) is attained either at $s_{12} = 0$ or at $s_{12} = s_{11}/2$. Then, since $s_{11} \le s_{22} \le s_{33}$, routine computation yields

$$\mathrm{dis}(Q) = \begin{cases} s_{11}s_{22}s_{33}\left(1 - \frac{s_{11}}{4s_{33}} - \frac{s_{22}}{4s_{33}}\right) \ge \frac{1}{2}s_{11}s_{22}s_{33}, & \text{if } s_{12} = 0, \\ s_{11}s_{22}s_{33}\left(1 - \frac{s_{11}}{4s_{22}} - \frac{s_{22}}{4s_{33}}\right) \ge \frac{1}{2}s_{11}s_{22}s_{33}, & \text{if } s_{12} = \frac{1}{2}s_{11}. \end{cases}$$

ii. $s_{23} = -s_{22}/2$. In this case, it follows from (2.16) that

$$\mathrm{dis}(Q) = s_{11}s_{22}s_{33} - s_{22}s_{12}s_{13} - \tfrac{1}{4}s_{11}(s_{22})^2 - s_{22}(s_{13})^2 - s_{33}(s_{12})^2. \tag{2.20}$$

Also, by (2.11.5),

$$2s_{13} \le s_{11} - 2s_{12}. \tag{2.21}$$

We calculate the derivative

$$\frac{\partial \mathrm{dis}(Q)}{\partial s_{13}} = -s_{22}(s_{12} + 2s_{13}) \le 0.$$

Hence, in order to minimize the value of $\mathrm{dis}(Q)$ we shall choose the maximum possible value for s_{13} under the conditions $2s_{13} \le s_{11}$ and (2.21), namely $s_{13} = s_{11}/2 - s_{12}$. Then, it follows from (2.20) that

$$\mathrm{dis}(Q) = a_3(s_{12})^2 + b_3 s_{12} + c_3,$$

where a_3, b_3, and c_3 are independent of s_{12}, and $a_3 = -s_{33} < 0$. Thus by a similar argument to the previous subcase we obtain

$$\mathrm{dis}(Q) \ge \tfrac{1}{2}s_{11}s_{22}s_{33}.$$

Case 2. $s_{23} \ge 0$. Then, the last condition in (2.11) follows from the others. Arguing as before, it follows that the minimal value of $\mathrm{dis}(Q)$ will be attained either at $s_{23} = 0$ or at $s_{23} = s_{22}/2$. Hence, by (2.16),

$$\mathrm{dis}(Q) \ge \min\{f(s_{12}, s_{13}, 0), f(s_{12}, s_{13}, s_{22}/2)\}$$
$$= s_{11}s_{22}s_{33} - s_{22}\left(\frac{s_{11}s_{22}}{4} - s_{12}s_{13}\right)$$
$$- s_{22}(s_{13})^2 - s_{33}(s_{12})^2.$$

Considering the right-hand side of this inequality as a function of s_{12} and s_{13}, and denoting it by $g(s_{12}, s_{13})$, simple analysis yields

$$\mathrm{dis}(Q) \ge \min_{\substack{0 \le s_{12} \le s_{11}/2, \\ 0 \le s_{13} \le s_{22}/2}} g(s_{12}, s_{13}) = g(s_{11}/2, s_{22}/2)$$
$$= s_{11}s_{22}s_{33}\left(1 - \frac{s_{11}}{4s_{22}} - \frac{s_{22}}{4s_{33}}\right)$$
$$\ge \tfrac{1}{2}s_{11}s_{22}s_{33}.$$

Hence (2.15) follows from Case 1 and Case 2. Also, tracing back through the above argument, it can be verified that every critical positive definite ternary quadratic form is equivalent to

$$Q(\mathbf{x}) = (x_1)^2 + (x_2)^2 + (x_3)^2 + x_1 x_2 + x_1 x_3 + x_2 x_3,$$

which implies the second part of our theorem. Theorem 2.1 is proved. □

2.3. Mordell's Inequality on Hermite's Constants and a Theorem of Korkin and Zolotarev

Lemma 2.4 (Mordell [1]).

$$\gamma_n \leq (\gamma_{n-1})^{\frac{n-1}{n-2}}.$$

Proof. Let

$$X_i = \sum_{j=1}^{n} a_{ij} x_j, \quad i = 1, 2, \ldots, n, \tag{2.22}$$

be n linear forms with real coefficients and determinant 1, and let b_{ij} be the cofactor of a_{ij} of A, the $n \times n$ matrix with entries a_{ij}. Define n linear forms $Y_1, Y_2, \ldots, Y_n$ in variables $y_1, y_2, \ldots, y_n$ by means of the identity

$$\sum_{i=1}^{n} X_i Y_i = \sum_{i=1}^{n} x_i y_i. \tag{2.23}$$

For example,

$$Y_i = \sum_{j=1}^{n} b_{ij} y_j, \quad i = 1, 2, \ldots, n.$$

Obviously, the n linear forms Y_i also have determinant 1.

Letting

$$Q(\mathbf{x}) = \mathbf{x} S \mathbf{x}' = \mathbf{x} A A' \mathbf{x}'$$

be a positive definite quadratic form with $\mathrm{dis}(Q) = 1$, we can write

$$Q(\mathbf{x}) = (X_1)^2 + (X_2)^2 + \cdots - (X_n)^2$$

with appropriate linear forms X_i satisfying (2.22). Then let

$$Q^*(\mathbf{y}) = (Y_1)^2 + (Y_2)^2 + \cdots + (Y_n)^2 \tag{2.24}$$

be the form associated with $Q(\mathbf{x})$ using (2.23). By applying a unimodular substitution in (2.24) and changing $Q^*(\mathbf{y})$ into an equivalent form, we may assume that the minimum $m(Q^*)$ is given by

$$(y_1, y_2, \ldots, y_n) = (0, 0, \ldots, 1).$$

In other words,

$$m(Q^*) = \sum_{i=1}^{n} (b_{in})^2. \tag{2.25}$$

In this case we must then also apply a corresponding unimodular substitution to the x_i's such that $\sum_{i=1}^{n} x_i y_i$ remains unaltered, and then $Q(\mathbf{x})$ is replaced by an equivalent form.

Now we estimate the values of $Q(\mathbf{z})$ for $z_n = 0$. In this case, $Q(\mathbf{z})$ becomes a quadratic form in $n-1$ variables, namely

$$Q^\star(\mathbf{z}) = \sum_{i=1}^{n} \left(\sum_{j=1}^{n-1} a_{ij} z_j \right)^2.$$

For convenience, we denote by $S^\star$ the $(n-1) \times (n-1)$ matrix with entries

$$s_{ij} = \sum_{k=1}^{n} a_{ki} a_{kj}$$

for $1 \le i \le n-1$ and $1 \le j \le n-1$, and denote by M the $n \times (n-1)$ matrix with entries a_{ij} for $1 \le i \le n$ and $1 \le j \le n-1$. Then, by (2.25), we have

$$\operatorname{dis}(Q^\star) = \det(S^\star) = \det(M'M) = \sum_{i=1}^{n} (b_{in})^2$$

$$= m(Q^*) \le \gamma_n,$$

and therefore

$$m(Q) \le m(Q^*) \le \gamma_{n-1} \operatorname{dis}(Q^\star)^{\frac{1}{n-1}} \le \gamma_{n-1} (\gamma_n)^{\frac{1}{n-1}}.$$

Since the right-hand side is independent of the form $Q(\mathbf{x})$, it follows from (2.3) that

$$\gamma_n \le \gamma_{n-1} (\gamma_n)^{\frac{1}{n-1}}$$

and consequently

$$\gamma_n \le (\gamma_{n-1})^{\frac{n-1}{n-2}}.$$

Lemma 2.4 is proved.   □

From (2.4), Lemma 2.4, Theorem 2.1, and their proofs it is easy to deduce the density of the densest lattice sphere packings in E^4, which was determined by Korkin and Zolotarev by a much more complicated method.

Theorem 2.2 (Korkin and Zolotarev [1]).

$$\delta^*(S_4) = \frac{\pi^2}{16}.$$

Also, the densest lattice packing of S_4 is unique up to rotation and reflection.

2.4. Perfect Forms, Voronoi's Method, and a Theorem of Korkin and Zolotarev

To determine the value of $\delta^*(S_n)$ by studying extreme forms, we introduce the concept of a *perfect form*. A positive definite quadratic form $Q(\mathbf{x})$ is called perfect if it is determined uniquely by the equations

$$Q(\mathbf{z}_i) = m(Q).$$

Extreme forms and perfect forms have the following important relationship.

Lemma 2.5 (Korkin and Zolotarev [2]). *Every extreme positive definite quadratic form is perfect.*

Proof. Let $Q(\mathbf{x}) = \mathbf{x}S\mathbf{x}'$ be an extreme positive definite quadratic form in n variables, and assume that

$$M(Q) = \{\pm\mathbf{z}_1, \ \pm\mathbf{z}_2, \ \ldots, \ \pm\mathbf{z}_k\}.$$

If, on the contrary, $Q(\mathbf{x})$ is not perfect, there is a nonzero symmetric matrix S^* such that

$$\mathbf{z}_i S^* \mathbf{z}_i' = 0, \quad i = 1, \ 2, \ \ldots, \ k.$$

Define

$$Q_\lambda(\mathbf{x}) = \mathbf{x}(S + \lambda S^*)\mathbf{x}'.$$

It is easy to see that for a suitable positive number α, $Q_\lambda(\mathbf{x})$ is positive definite and

$$m(Q_\lambda) = m(Q) = 1 \tag{2.26}$$

whenever $|\lambda| \leq \alpha$. It is well-known from linear algebra that there is an orthogonal matrix U such that both USU' and US^*U' are diagonal, with diagonal entries s_i and s_i^*, respectively. Hence,

$$\text{dis}(Q_\lambda) = \prod_{i=1}^{n}(s_i + \lambda s_i^*).$$

Writing

$$f(\lambda) = \log \text{dis}(Q_\lambda),$$

simple analysis yields that $f(\lambda)$ is a strictly concave function of λ. There-fore, one has

$$\text{dis}(Q_\lambda) < \text{dis}(Q) \tag{2.27}$$

for $\lambda > 0$, or for $\lambda < 0$. Then, (2.26) and (2.27) together yield,

$$\frac{m(Q_\lambda)}{\sqrt[n]{\text{dis}(Q_\lambda)}} > \frac{m(Q)}{\sqrt[n]{\text{dis}(Q)}},$$

which contradicts the assumption that $Q(\mathbf{x})$ is extreme. Lemma 2.5 is proved.   □

Using Lemma 2.5, we can determine all the extreme forms if we can determine all the perfect forms. For this reason, since a perfect form can be determined by its minimum integer points, Voronoi [1] developed the following method. The method is very elegant. However, its proof is very complicated. So, here we only introduce the method itself.

Voronoi's Method. For convenience, we write

$$d = \frac{n(n+1)}{2}.$$

Let $Q_1(\mathbf{x})$ be a perfect form in n variables with $m(Q_1) = 1$ and assume that

$$M(Q_1) = \{\pm\mathbf{z}_1,\ \pm\mathbf{z}_2,\ \ldots,\ \pm\mathbf{z}_k\},$$

where $\mathbf{z}_i = (z_{i1},\ z_{i2},\ \ldots,\ z_{in})$. Then, corresponding to $\mathbf{z}_i$ and $Q_1(\mathbf{x})$, respectively, in E^d we introduce a point

$$\mathbf{z}_i^* = (z_{i1}z_{i1}, z_{i1}z_{i2}, \ldots, z_{i1}z_{in}, z_{i2}z_{i2}, z_{i2}z_{i3}, \ldots, z_{i2}z_{in}, \ldots, z_{in}z_{in})$$

and a *polyhedral convex cone*

$$V_1 = \left\{ \sum_{i=1}^{k} \alpha_i \mathbf{z}_i^* :\ \alpha_i \geq 0 \right\}.$$

Assume that V_1 has $l_1 - 1$ *facets* F_i, $i = 2, 3, \ldots, l_1$, which are determined by

$$\left\{ \mathbf{o}^*, \mathbf{z}_{i_1}^*, \mathbf{z}_{i_2}^*, \ldots, \mathbf{z}_{i_{f(i)}}^* \right\}.$$

Let $Q_i^*(\mathbf{x})$ be nonzero quadratic forms in n variables such that

$$Q_i^*(\mathbf{z}_{i_j}) = 0, \quad j = 1, 2, \ldots, f(i).$$

Defining

$$\lambda_i = \min_{\mathbf{z} \in Z_n,\, Q_i^*(\mathbf{z}) < 0} \frac{Q_1(\mathbf{z}) - 1}{-Q_i^*(\mathbf{z})}$$

and

$$Q_i(\mathbf{x}) = Q_1(\mathbf{x}) + \lambda_i Q_i^*(\mathbf{x})$$

for $i = 2, 3, \ldots, l_1$, then $Q_i(\mathbf{x})$ are new perfect forms such that $m(Q_i) = 1$ and the corresponding polyhedral cone V_i joins V_1 at the whole *facet* F_i. For convenience, we say that $Q_i(\mathbf{x})$ is a neighbor of $Q_1(\mathbf{x})$.

Repeating this process for $Q_2(\mathbf{x})$, $Q_3(\mathbf{x})$, $\ldots$, $Q_{l_1}(\mathbf{x})$ and their neighbors, a sequence of perfect forms

$$Q_1(\mathbf{x}),\ Q_2(\mathbf{x}),\ \ldots,\ Q_l(\mathbf{x}) \tag{2.28}$$

can be found such that every neighbor of any of these forms is equivalent to one of them. Then, (2.28) is a complete system of perfect positive definite quadratic forms in n variables. In other words, every perfect form in n variables is equivalent to one of the l forms in (2.28).

By this method, one can prove the following result with routine argument.

Lemma 2.6 (Korkin and Zolotarev [3]). *Let*

$$U_n(\mathbf{x}) = \sum_{1 \le i \le j \le n} x_i x_j, \quad n \ge 2,$$

$$V_n(\mathbf{x}) = U_n(\mathbf{x}) - x_1 x_2, \quad n \ge 4,$$

and

$$W_5(\mathbf{x}) = \sum_{i=1}^{5} (x_i)^2 - \frac{1}{2} \sum_{i=2}^{5} x_1 x_i + \frac{1}{2} \sum_{2 \le i < j \le 4} x_i x_j - \sum_{i=2}^{4} x_i x_5.$$

For $n \le 5$, every perfect positive definite quadratic form $Q(\mathbf{x})$ with $m(Q) = 1$ is equivalent to one of the seven forms $U_2(\mathbf{x})$, $U_3(\mathbf{x})$, $U_4(\mathbf{x})$, $V_4(\mathbf{x})$, $U_5(\mathbf{x})$, $V_5(\mathbf{x})$, or $W_5(\mathbf{x})$.

Then, the density of the densest lattice sphere packings in E^5 can be deduced from (2.3), (2.4), Lemma 2.5, and Lemma 2.6.

Theorem 2.3 (Korkin and Zolotarev [3]).

$$\delta^*(S_5) = \frac{\pi^2}{15\sqrt{2}}.$$

In addition, the densest lattice packing of S_5 is unique up to rotation and reflection.

2.5. The Korkin-Zolotarev Reduction and Theorems of Blichfeldt, Barnes, and Vetčinkin

Definition 2.2. *A positive definite quadratic form $Q(\mathbf{x})$ is said to be K-Z reduced if*

$$Q(\mathbf{x}) = \sum_{i=1}^{n} c_i \left(x_i + \sum_{j=i+1}^{n} t_{ij} x_j \right)^2,$$

where $|t_{ij}| \leq \frac{1}{2}$ and

$$c_i = \min_{(z_i, z_{i+1}, \ldots, z_n) \in Z_{n-i+1} \setminus \{0\}} \left\{ \sum_{j=i}^{n} c_j \left(z_j + \sum_{k=j+1}^{n} t_{jk} z_k \right)^2 \right\}.$$

Lemma 2.7 (Korkin and Zolotarev [2]). *Every positive definite quadratic form is equivalent to a K-Z reduced form.*

Proof. Let $Q(\mathbf{x}) = \mathbf{x}S\mathbf{x}'$ be a positive definite quadratic form in n variables, let A be an $n \times n$ matrix such that $S = AA'$, and let Λ be the lattice $\{\mathbf{z}A : \mathbf{z} \in Z_n\}$. Now we proceed to show that the lattice Λ has a special basis $\{\mathbf{b}_1, \mathbf{b}_2, \ldots, \mathbf{b}_n\}$ that corresponds to a K-Z reduced form.

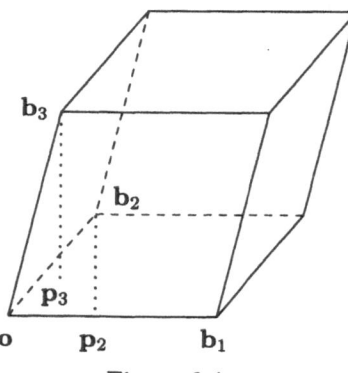

Figure 2.1

To seek such a basis efficiently, we denote the i-dimensional space generated by $\{\mathbf{b}_1, \mathbf{b}_2, \ldots, \mathbf{b}_i\}$ by H_i, the parallelepiped $\{\sum_{j=1}^{i} \lambda_j \mathbf{b}_j : |\lambda_j| \leq \frac{1}{2}\}$ by P_i, the orthogonal projection of $\mathbf{b}_i$ to H_{i-1} by $\mathbf{p}_i$ (see Figure 2.1), and, as usual, the distance between a point $\mathbf{x}$ and a set X by $d(\mathbf{x}, X)$. Writing

$$\mathbf{q}_i = \mathbf{b}_i - \mathbf{p}_i \qquad (2.29)$$

and

$$P_i^* = \left\{ \sum_{j=1}^{i} \lambda_j \mathbf{q}_j : |\lambda_j| \leq \frac{1}{2} \right\},$$

it is easy to see that
$$v_i(P_i) = v_i(P_i^*),\tag{2.30}$$

where $v_i(X)$ indicates the i-dimensional volume of X.

First, take $\mathbf{b}_1$ to be a point of the lattice Λ such that

$$\langle \mathbf{b}_1, \mathbf{b}_1 \rangle = \min_{\mathbf{a} \in \Lambda \setminus \{\mathbf{o}\}} \langle \mathbf{a}, \mathbf{a} \rangle.$$

Then, choose $\mathbf{b}_2$ to be a point of Λ such that

$$d(\mathbf{b}_2, H_1) = \min_{\mathbf{a} \in \Lambda \setminus H_1} d(\mathbf{a}, H_1)$$

and

$$\mathbf{p}_2 \in P_1^*.\tag{2.31}$$

Inductively, if $\mathbf{b}_1$, $\mathbf{b}_2$, ..., $\mathbf{b}_i$ have been fixed, we choose $\mathbf{b}_{i+1}$ to be a point of Λ such that

$$d(\mathbf{b}_{i+1}, H_i) = \min_{\mathbf{a} \in \Lambda \setminus H_i} d(\mathbf{a}, H_i)\tag{2.32}$$

and

$$\mathbf{p}_{i+1} \in P_i^*\tag{2.33}$$

((2.31) and (2.33) are guaranteed by the periodic behavior of the *sublattice* $\{\sum_{j=1}^{i} z_j \mathbf{b}_j : z_j \in Z\}$ and (2.30)). Thus, we obtain a set of n points $\{\mathbf{b}_1, \mathbf{b}_2, ..., \mathbf{b}_n\}$ that is a basis for Λ and a corresponding set of n points $\{\mathbf{q}_1, \mathbf{q}_2, ..., \mathbf{q}_n\}$ that is an orthogonal basis for E^n. Therefore, every point $\mathbf{x}$ of E^n can be represented as

$$\mathbf{x} = \sum_{i=1}^{n} x_i \mathbf{b}_i = \sum_{i=1}^{n} y_i \mathbf{q}_i,\tag{2.34}$$

where, by (2.29) and (2.33),

$$y_i = x_i + \sum_{j=i+1}^{n} t_{ij} x_j\tag{2.35}$$

and

$$|t_{ij}| \le \tfrac{1}{2}.\tag{2.36}$$

Since $\{\mathbf{q}_1, \mathbf{q}_2, ..., \mathbf{q}_n\}$ is an orthogonal basis of E^n, by (2.34) and (2.35) the form $Q(\mathbf{x})$ is equivalent to

$$Q^*(\mathbf{x}) = \left\langle \sum_{i=1}^{n} x_i \mathbf{b}_i, \sum_{i=1}^{n} x_i \mathbf{b}_i \right\rangle = \left\langle \sum_{i=1}^{n} y_i \mathbf{q}_i, \sum_{i=1}^{n} y_i \mathbf{q}_i \right\rangle$$

$$= \sum_{i=1}^{n} \langle \mathbf{q}_i, \mathbf{q}_i \rangle (y_i)^2 = \sum_{i=1}^{n} \langle \mathbf{q}_i, \mathbf{q}_i \rangle \left(x_i + \sum_{j=i+1}^{n} t_{ij} x_j \right)^2.$$

By (2.36) and (2.32) it is easy to see that $Q^*(\mathbf{x})$ is K-Z reduced with $c_i = \langle \mathbf{q}_i, \mathbf{q}_i \rangle$. Lemma 2.7 is proved. $\square$

If $Q(\mathbf{x})$ is K-Z reduced, then

$$m(Q) = c_1 \quad \text{and} \quad \text{dis}(Q) = c_1 c_2 \cdots c_n.$$

Thus, letting $\mathcal{R}$ denote the family of K-Z reduced positive definite quadratic forms in n variables, by (2.1), (2.3), (2.5), and Lemma 2.7 it follows that

$$\delta^*(S_n) = \sup_{Q(\mathbf{x}) \in \mathcal{R}} \frac{\omega_n(c_1)^{n/2}}{2^n \sqrt{c_1 c_2 \cdots c_n}} \tag{2.37}$$

and

$$\gamma_n = \sup_{Q(\mathbf{x}) \in \mathcal{R}} \frac{c_1}{\sqrt[n]{c_1 c_2 \cdots c_n}}. \tag{2.38}$$

Therefore, the values of $\delta^*(S_n)$ and γ_n can be determined by studying the relationship between c_1 and $\sqrt[n]{c_1 c_2 \cdots c_n}$ for the K-Z reduced positive definite quadratic forms in n variables.

Lemma 2.8 (Korkin and Zolotarev [2]). *The outer coefficients of a K-Z reduced positive definite quadratic form satisfy*

$$c_{i+1} \geq \tfrac{3}{4} c_i \quad for \ i = 1, 2, \ldots, n-1, \tag{2.39}$$

and

$$c_{i+2} \geq \tfrac{2}{3} c_i \quad for \ i = 1, 2, \ldots, n-2. \tag{2.40}$$

Proof. For convenience, we assume that $Q(\mathbf{x})$ is a K-Z reduced ternary form with $c_1 = 1$. By Definition 2.2, taking $\mathbf{x} = (0,1,0)$, $(0,0,1)$, $(-1,1,1)$, and $(0,1,-1)$, respectively, it follows that

$$c_2 + (t_{12})^2 \geq 1, \tag{2.41}$$

$$c_3 + c_2 (t_{23})^2 + (t_{13})^2 \geq 1, \tag{2.42}$$

$$c_3 + c_2 (1 + t_{23})^2 + (1 - t_{12} - t_{13})^2 \geq 1, \tag{2.43}$$

and

$$c_3 + c_2 (1 - t_{23})^2 + (t_{12} - t_{13})^2 \geq 1. \tag{2.44}$$

Similarly, taking $\mathbf{x} = (0,0,1)$ and comparing $Q(\mathbf{x})$ with c_2, one has

$$c_3 + c_2 (t_{23})^2 \geq c_2. \tag{2.45}$$

First, since $|t_{ij}| \leq \tfrac{1}{2}$, it follows from (2.41) that

$$c_2 \geq 1 - (t_{12})^2 \geq \tfrac{3}{4}, \tag{2.46}$$

which implies (2.39). Now we proceed to prove (2.40) by dealing with three cases.

Case 1. $|t_{23}| \leq \frac{1}{3}$. Then, (2.45) and (2.46) imply

$$c_3 \geq c_2 \left(1 - (t_{23})^2\right) \geq \frac{2}{3}.$$

Case 2. $-\frac{1}{2} \leq t_{23} < -\frac{1}{3}$. By (2.41) and (2.45) it follows that

$$c_3 \geq c_2 \left(1 - (t_{23})^2\right) \geq \left(1 - (t_{12})^2\right)\left(1 - (t_{23})^2\right)$$

and therefore

$$(t_{12})^2 \geq 1 - \frac{c_3}{1 - (t_{23})^2}. \tag{2.47}$$

Similarly, by (2.42) and (2.45) it follows that

$$(t_{13})^2 \geq 1 - \frac{c_3}{1 - (t_{23})^2}. \tag{2.48}$$

Without loss of generality, we assume that both t_{12} and t_{13} are positive. Hence, by (2.43), (2.45), (2.47), and (2.48) we have

$$c_3 + \frac{1 + t_{23}}{1 - t_{23}}c_3 + \left(1 - 2\sqrt{1 - \frac{c_3}{1 - (t_{23})^2}}\right)^2 \geq 1.$$

Then, routine analysis yields $c_3 \geq \frac{2}{3}$.

Case 3. $\frac{1}{3} < t_{23} \leq \frac{1}{2}$. As in Case 2, by (2.44), (2.45), (2.36), and (2.48) we have

$$c_3 + \frac{1 - t_{23}}{1 + t_{23}}c_3 + \left(\frac{1}{2} - \sqrt{1 - \frac{c_3}{1 - (t_{23})^2}}\right)^2 \geq 1.$$

Then, routine analysis yields $c_3 \geq \frac{2}{3}$.

Thus, (2.40) is proved and hence the proof is complete. $\square$

Remark 2.2. *The values of $\delta^*(S_2)$, $\delta^*(S_3)$, and $\delta^*(S_4)$ can be directly deduced from (2.37) and Lemma 2.8. Thus, we obtain different proofs for Theorem 2.1 and Theorem 2.2. Similarly, one can determine the values of γ_2, γ_3, and γ_4 by (2.38) and Lemma 2.8.*

Besides Lemma 2.8, there is another basic result about the outer coefficients of a K-Z reduced positive definite quadratic form in n variables.

Lemma 2.9 (Vetčinkin [2]).

$$c_{i+1}c_{i+2}\cdots c_{i+j} \geq \frac{(c_i)^j}{\sqrt[j+1]{\gamma_{j+1}}},$$

where $i + j \leq n$.

Proof. Assume that

$$Q(\mathbf{x}) = \sum_{k=1}^{n} c_k \left(x_k + \sum_{l=k+1}^{n} t_{kl} x_l \right)^2$$

is K-Z reduced. Let

$$Q_{ij}(\mathbf{x}) = \sum_{k=i}^{i+j} c_k \left(x_k + \sum_{l=k+1}^{i+j} t_{kl} x_l \right)^2.$$

Then $Q_{ij}(\mathbf{x})$ is a positive definite quadratic form in $j+1$ variables satisfying

$$m(Q_{ij}) = c_i$$

and

$$\mathrm{dis}(Q_{ij}) = c_i c_{i+1} \cdots c_{i+j}.$$

Then, by the definition of Hermite's constant it follows that

$$\frac{c_i}{\sqrt[j+1]{c_i c_{i+1} \cdots c_{i+j}}} = \frac{m(Q_{ij})}{\sqrt[j+1]{\mathrm{dis}(Q_{ij})}} \leq \gamma_{j+1},$$

which proves Lemma 2.9. ☐

Using (2.37), Lemma 2.7, Lemma 2.8, and Lemma 2.9, one can determine the densities of the densest lattice sphere packings in E^6, E^7, and E^8 by considering many different cases. Since it is impossible to relate the details in a few pages, we simply quote the results here.

Theorem 2.4 (Blichfeldt [2]).

$$\delta^*(S_6) = \frac{\pi^3}{48\sqrt{3}}, \quad \delta^*(S_7) = \frac{\pi^3}{105}, \quad and \quad \delta^*(S_8) = \frac{\pi^4}{384}.$$

Theorem 2.5 (Barnes [4] and Vetčinkin [2]). *The densest lattice packing of S_n, $n = 6$, 7, or 8, is unique up to rotation and reflection.*

Remark 2.3. *By Lemma 2.4, one can deduce the value of $\delta^*(S_8)$ from that of $\delta^*(S_7)$.*

Remark 2.4. *Besides L-S-M reduction and K-Z reduction, there are several others such as Hermite's reduction and Selling's reduction that have been studied by Afflerbach, Babai, Baranovskii, Delone, Lenstra, Lenstra Jr., Lovász, Ryškov, Štogrin, Tammela, Venkov, van der Waerden, and many others. Unfortunately, when n is comparatively large, none of these reductions are efficient enough to deal with either $\delta^*(S_n)$ or $k^*(S_n)$.*

2.6. Perfect Forms, the Lattice Kissing Numbers of Spheres, and Watson's Theorem

To determine the lattice kissing numbers of spheres, we introduce another important result concerning perfect forms.

Lemma 2.10 (Voronoi [2]). *For every positive definite quadratic form* $Q(\mathbf{x})$ *there is a perfect form* $Q^*(\mathbf{x})$ *such that*

$$M(Q) \subseteq M(Q^*).$$

Proof. Let $Q(\mathbf{x}) = \mathbf{x}S\mathbf{x}'$ be an imperfect positive definite quadratic form in n variables and assume that

$$M(Q) = \{\pm\mathbf{z}_1, \ \pm\mathbf{z}_2, \ \ldots, \ \pm\mathbf{z}_k\}\,.$$

Since $Q(\mathbf{x})$ is imperfect, there is a nonzero quadratic form $Q^*(\mathbf{x}) = \mathbf{x}S^*\mathbf{x}'$ such that

$$Q^*(\mathbf{z}_i) = \mathbf{z}_i S^* \mathbf{z}_i' = 0, \quad i = 1, \ 2, \ \ldots, \ k.$$

Let $\mathcal{Q}_1$ be the family of all positive definite quadratic forms

$$Q_\lambda(\mathbf{x}) = Q(\mathbf{x}) + \lambda Q^*(\mathbf{x})$$

such that $m(Q_\lambda) = m(Q)$. For convenience, we write

$$Z^* = Z_n \setminus \{M(Q) \cup \{\mathbf{o}\}\}.$$

Clearly, we have

$$\min_{Q_\lambda \in \mathcal{Q}_1} \left\{ \min_{\mathbf{z} \in Z^*} Q_\lambda(\mathbf{z}) \right\} = m(Q).$$

Thus, choosing a form $Q_1(\mathbf{x}) \in \mathcal{Q}_1$ such that

$$\min_{\mathbf{z} \in Z^*} Q_1(\mathbf{z}) = m(Q),$$

it is obvious that

$$M(Q) \subset M(Q_1).$$

If $Q_1(\mathbf{x})$ is perfect, the assertion of Lemma 2.10 is proved by taking $Q^*(\mathbf{x}) = Q_1(\mathbf{x})$. Otherwise, replacing $Q(\mathbf{x})$ by $Q_1(\mathbf{x})$ (or its successors) and repeating this process, one can obtain a sequence of positive definite quadratic forms

$$Q(\mathbf{x}), \ Q_1(\mathbf{x}), \ Q_2(\mathbf{x}), \ \ldots \qquad (2.49)$$

such that

$$m(Q_i) = m(Q), \quad i = 1, \ 2, \ \ldots,$$

and

$$M(Q) \subset M(Q_1) \subset M(Q_2) \subset \cdots. \qquad (2.50)$$

On the other hand, since $k^*(S_n)$ is bounded from both sides, the sequence (2.49) terminates at a certain form, say $Q_l(\mathbf{x})$. This form is perfect. Hence, taking $Q^*(\mathbf{x}) = Q_l(\mathbf{x})$, Lemma 2.10 is proved. $\square$

By (2.2) and this lemma, we can determine the value of $k^*(S_n)$ if we can find all perfect forms in n variables. However, complete sets of positive definite perfect forms are known only when $n \leq 6$ (see Lemma 2.6 and Barnes [4]). Using Lemma 2.10 and considering many complicated cases, G.L. Watson was able to prove the following theorem.

Theorem 2.6 (Watson [4]).

n	4	5	6	7	8	9
$k^*(S_n)$	24	40	72	126	240	272

In addition, in each of these dimensions, the lattice packing of S_n at which $k^*(S_n)$ is attained is unique up to rotation and reflection.

Remark 2.5. *Let D_n and E_n be the lattices defined in Chapter 9 and let*

$$\Lambda_n = \begin{cases} D_n & if\ 3 \leq n \leq 5, \\ E_n & if\ 6 \leq n \leq 8. \end{cases}$$

It is easy to verify that both $\delta^*(S_n)$ and $k^*(S_n)$ can be realized at $S_n + \Lambda_n$ when $3 \leq n \leq 8$. In E^9, fix a lattice Λ_8 in the plane

$$H = \{\mathbf{x} \in E^9 : x_9 = 0\}$$

and take

$$\Lambda_9 = \{z\mathbf{u} + \Lambda_8 : z \in Z\}$$

such that $S_9 + \Lambda_9$ is the densest lattice packing of this type. Then $k^*(S_9)$ can be realized in $S_9 + \Lambda_9$. At present we do not know the exact values of $\delta^*(S_n)$ for any $n \geq 9$, nor the exact values of $k^*(S_n)$ for any $n \geq 10$, except 24, which will be discussed in Chapter 9.

2.7. Three Mathematical Geniuses: Zolotarev, Minkowski, and Voronoi

In order to study a subject that has developed slowly over the centuries it is important to know something of its history and its major contributors.

Many great mathematicians, such as Blichfeldt, Delone, Dirichlet, Gauss, Hermite, Korkin, Lagrange, Minkowski, Voronoi, Watson, and Zolotarev, have made important contributions to our knowledge of lattice sphere packings or, alternatively, positive definite quadratic forms. In this section we will briefly introduce three of them, Zolotarev, Minkowski, and Voronoi. For more details about them we refer to Ozhigova [1].

Egor Ivanovich Zolotarev was born in St. Petersburg in 1847, the son of a watch store owner. At the age of seventeen he entered the Mathematical-Physical Faculty of St. Petersburg University, where he attended the lectures of P.L. Chebyshev and A.N. Korkin, with whom he later collaborated on a number of important papers. In 1867 he completed his undergraduate studies and earned the degree of candidate. In 1874 he obtained a doctorate, also from St. Petersburg University, with a dissertation *The Theory of Whole Complex Numbers with Application to the Integral Calculus*. This remarkable work made him one of the most prominent number theorists of his time.

During the summers of 1872 and 1876 the university sent Zolotarev abroad for four months. This gave him the oppourtunity to visit Berlin, Heidelberg, and Paris, to attend lectures of Kirchhoff, Kummer, and Weierstrass, and to meet Hermite. In 1876, Zolotarev was promoted to professor at St. Petersburg University. In the very same year he was elected a member of the St. Petersburg Academy of Sciences.

In the summer of 1878, while at his most creative and productive, Zolotarev's life came to an abrupt and tragic end. On July 2, while preparing to visit relatives in their summer home, he was knocked down by a vehicle and died of blood poisoning on July 19.

Besides his joint work with Korkin on positive definite quadratic forms, some of which was discussed in the previous sections, Zolotarev made important contributions to the arithmetic of algebraic numbers, the theory of algebraic functions, and the theory of best approximation of functions by polynomials. His ideas and methods have had a fundamental influence on the further development of these subjects.

Hermann Minkowski, one of the most remarkable geniuses in the history of mathematics, was born in the town of Aleksoty in Russia (now in Lithuania) in 1864. While he was still a child his family moved to Königsberg. At the age of fifteen he graduated from the local gymnasium and then studied at the Universities of Königsberg and Berlin, attending the lectures of Helmholtz, Hurwitz, Kirchhoff, Kronecker, Kummer, Lindemann, Weber, Weierstrass, and others. In 1882, while he was still a student, Minkowski took part in a Paris Academy of Sciences competition on the problem of representing an integer as a sum of five squares. His paper made a profound impression on the jury, which included Bertrand, Bonnet, Bouquet,

Hermite, and Jordan. The judges were so impressed that the author was forgiven for writing his paper in German, which was against the rules of the Academy. As a result, in 1883, Minkowski and Smith, a well-known number theorist at Oxford University, shared the Grand Prix des Sciences Mathématiques of the Paris Academy of Sciences.

In 1885, H. Minkowski obtained a doctorate from the University of Königsberg with a dissertation on positive definite quadratic forms in n variables. He then began teaching at Königsberg, and in 1887 he moved to Bonn. In 1892 he became an associate, and in 1894 a full professor. In 1895 he was promoted to the chair of Hilbert, a student colleague and friend of his at Königsberg. He stayed for only two years and then accepted an offer from ETH-Zürich. In 1902, due to the efforts of Hilbert and Klein, a new chair for mathematics and physics was created at the University of Göttingen. Minkowski took the new chair from 1902 until his sudden death in 1909.

Minkowski is known as the founder of the geometry of numbers. In order to investigate positive definite quadratic forms, he tried to represent the problems in geometric terms, and he then discovered the fundamental principles of geometry of numbers. Minkowski also made important contributions to several other subjects, such as convex geometry and even relativity. There are many mathematical terms such as *Minkowski's theorem*, the *Minkowski-Hlawka theorem*, the *Brunn-Minkowski inequality*, and the *Minkowski metric* that bear his name in memory of his great contributions.

Georgii Feodosevich Voronoi was born in 1868 to the family of a professor of Russian literature at the Lyceum in Nezhinsk. After graduating from the gymnasium in Priluki in 1885, Voronoi entered the Mathematical-Physical Faculty of St. Petersburg University to study mathematics. When he was still a student at the gymnasium he wrote his first mathematical article, which was published in 1885 in the *Journal of Elementary Mathematics*.

During his university years, Voronoi studied the properties of *Bernoulli numbers* and made a discovery that attracted the attention of A. Korkin and A.A. Markov. He obtained his master's degree in 1894 with a dissertation *On Integral Algebraic Numbers Depending on the Root of an Equation of the Third Degree* and a doctorate in 1896 with a dissertation *On a Generalization of the Algorithm of Continued Fractions*, both from St. Petersburg. For these dissertations he was awarded a Bunyakovskii prize in 1896 by the St. Petersburg Academy of Sciences.

In 1894, Voronoi became a professor at the University of Warsaw. In 1907, he was elected corresponding member of the St. Petersburg Academy of Sciences. Unfortunately, he died in 1908, at the age of forty.

In 1908 and 1909, Voronoi published three long papers (altogether 287 pages) on positive definite quadratic forms in the *Journal für die reine und angewandte Mathematik*. These are three of the most important papers in

this subject and have had a profound influence on its further development
in this century, especially on the work of Baranovskii, Barnes, Delone,
Ryškov, Štogrin, Tammela, Venkov, Watson, and others. He also made
contributions to analytic number theory, for example to the divisor problem
and to the theory of the *Riemann zeta function*.

3. Lower Bounds for the Packing Densities of Spheres

3.1. The Minkowski-Hlawka Theorem

While it is impossible to determine the values of $\delta^*(S_n)$ for all dimensions, obtaining reasonable asymptotic bounds for them turns out to be both important and interesting. In 1905, by studying positive definite quadratic forms, Minkowski [6] proved

$$\delta^*(S_n) \geq \frac{\zeta(n)}{2^{n-1}},$$

where $\zeta(x) = \sum_{k=1}^{\infty} 1/k^x$ is the Riemann zeta function, and made a general conjecture for bounded *star bodies*. Forty years later his conjecture was proved by Hlawka [1]. In this section we prove the Minkowski-Hlawka theorem for centrally symmetric convex bodies.

Theorem 3.1 (The Minkowski-Hlawka Theorem). *Let C be an n-dimensional centrally symmetric convex body. Then*

$$\delta^*(C) \geq \frac{\zeta(n)}{2^{n-1}}.$$

To prove this theorem we introduce two general lemmas.

Lemma 3.1 (Davenport and Rogers [1]). *Let $f(\mathbf{x})$ be a continuous function vanishing outside a bounded region, and for a real number λ, write*

$$I(\lambda) = \int_{-\infty}^{\infty} \cdots \int_{-\infty}^{\infty} f(\mathbf{x}) dx_1 \cdots dx_{n-1},$$

where $\mathbf{x} = (x_1, \ldots, x_{n-1}, \lambda)$. *Let* Λ^* *be a lattice in the* $(n-1)$*-dimensional subspace* $x_n = 0$ *and let* α *be a fixed positive number. Then there exist a point* $\mathbf{y} = (y_1, \ldots, y_{n-1}, \alpha)$ *and a lattice* $\Lambda = \{z\mathbf{y} + \Lambda^* : z \in Z\}$ *such that*

$$\sum_{\mathbf{u} \in \Lambda,\, u_n \neq 0} f(\mathbf{u}) \leq \frac{1}{\det(\Lambda^*)} \sum_{k \in Z \setminus \{0\}} I(k\alpha). \tag{3.1}$$

Proof. Without loss of generality, we may assume that $\Lambda^* = Z_{n-1}$. Then

$$\sum_{\mathbf{u} \in \Lambda,\, u_n \neq 0} f(\mathbf{u}) = \sum_{k \in Z \setminus \{0\}} \sum_{\mathbf{z} \in \Lambda^*} f(\mathbf{z} + k\mathbf{y}).$$

For convenience we write $\mathbf{x} = (x_1, x_2, \ldots, x_{n-1}, k\alpha)$ and assume $k > 0$. Then, averaging the inner sum on the right-hand side with respect to $(y_1, y_2, \ldots, y_{n-1})$ and substituting $ky_i = x_i$, we have

$$\int_0^1 \cdots \int_0^1 \left\{ \sum_{\mathbf{z} \in \Lambda^*} f(\mathbf{z} + k\mathbf{y}) \right\} dy_1 \cdots dy_{n-1}$$

$$= \frac{1}{k^{n-1}} \int_0^k \cdots \int_0^k \left\{ \sum_{\mathbf{z} \in \Lambda^*} f(\mathbf{z} + \mathbf{x}) \right\} dx_1 \cdots dx_{n-1}$$

$$= \frac{1}{k^{n-1}} \sum_{\substack{\mathbf{z}' \in \Lambda^* \\ 0 \leq z_i' \leq k-1}} \int_0^1 \cdots \int_0^1 \left\{ \sum_{\mathbf{z} \in \Lambda^*} f(\mathbf{z} + \mathbf{z}' + \mathbf{x}) \right\} dx_1 \cdots dx_{n-1}$$

$$= \int_0^1 \cdots \int_0^1 \left\{ \sum_{\mathbf{z} \in \Lambda^*} f(\mathbf{z} + \mathbf{x}) \right\} dx_1 \cdots dx_{n-1}$$

$$= \int_{-\infty}^{\infty} \cdots \int_{-\infty}^{\infty} f(\mathbf{x}) dx_1 \cdots dx_{n-1} = I(k\alpha).$$

Therefore,

$$\int_0^1 \cdots \int_0^1 \left\{ \sum_{\mathbf{u} \in \Lambda,\, u_n \neq 0} f(\mathbf{u}) \right\} dy_1 \cdots dy_{n-1} = \sum_{k \in Z \setminus \{0\}} I(k\alpha).$$

Then there is a point $\mathbf{y}$ such that (3.1) holds with $\det(\Lambda^*) = 1$, which proves Lemma 3.1.   □

Lemma 3.2 (Hlawka [1]). *Let* $f(\mathbf{x})$ *be a bounded Riemann integrable function vanishing outside a bounded region, and let* ϵ *be a positive number. Then, there exists a lattice* Λ *with determinant 1 such that*

$$\sum_{\mathbf{u} \in \Lambda \setminus \{o\}} f(\mathbf{u}) < \int_{E^n} f(\mathbf{x}) d\mathbf{x} + \epsilon.$$

Proof. Without loss of generality, we may assume that $f(\mathbf{x})$ is continuous. Let α and β be positive numbers such that

$$\beta = \alpha^{1/(1-n)},$$

and denote by Λ^* the lattice βZ_{n-1}, in the subspace $x_n = 0$. Then

$$\det(\Lambda^*) = \alpha^{-1}. \tag{3.2}$$

Since $f(\mathbf{x}) = 0$, it follows that if $\mathbf{x}$ is outside a bounded region,

$$f(\mathbf{u}) = 0 \tag{3.3}$$

holds for every point $\mathbf{u} \in \Lambda^* \setminus \{\mathbf{o}\}$ when α is sufficiently small. Also, by the definition of $I(\lambda)$ in the previous lemma it follows that

$$\lim_{\alpha \to 0} \sum_{k \in Z \setminus \{0\}} \alpha I(k\alpha) = \int_{E^n} f(\mathbf{x}) d\mathbf{x}. \tag{3.4}$$

Thus, by (3.2), (3.3), (3.4), and Lemma 3.1, for any $\epsilon > 0$ there exist a suitable number α and a suitable point $\mathbf{y} = (y_1, \ldots, y_{n-1}, \alpha)$ such that the corresponding lattice,

$$\Lambda = \{z\mathbf{y} + \Lambda^* : z \in Z\},$$

satisfies both $\det(\Lambda) = 1$ and

$$\sum_{\mathbf{u} \in \Lambda \setminus \{\mathbf{o}\}} f(\mathbf{u}) = \sum_{\mathbf{u} \in \Lambda, \, u_n \neq 0} f(\mathbf{u}) \leq \sum_{k \in Z \setminus \{0\}} \alpha I(k\alpha)$$

$$< \int_{E^n} f(\mathbf{x}) d\mathbf{x} + \epsilon.$$

Lemma 3.2 is proved. □

Proof of Theorem 3.1. It is sufficient to show that if

$$v(C) < 2\zeta(n), \tag{3.5}$$

then there is a lattice Λ with determinant 1 such that $\frac{1}{2}C + \Lambda$ is a packing. In other words,

$$C \cap \Lambda = \{\mathbf{o}\}. \tag{3.6}$$

Let $\chi(\mathbf{x})$ be the characteristic function of C and define

$$f(\mathbf{x}) = \sum_{k=1}^{\infty} \mu(k) \chi(k\mathbf{x}). \tag{3.7}$$

where $\mu(k)$ is the *Möbius function*. It is well known that

$$\sum_{k=1}^{\infty} \frac{\mu(k)}{k^n} = \frac{1}{\zeta(n)} \tag{3.8}$$

and

$$\sum_{k|m} \mu(k) = \begin{cases} 1 & \text{if } m = 1, \\ 0 & \text{otherwise.} \end{cases} \tag{3.9}$$

It follows from (3.7), (3.8), and (3.5) that

$$\begin{aligned}
\int_{E^n} f(\mathbf{x})d\mathbf{x} &= \sum_{k=1}^{\infty} \mu(k) \int_{E^n} \chi(k\mathbf{x})d\mathbf{x} \\
&= \sum_{k=1}^{\infty} \frac{\mu(k)}{k^n} v(C) \\
&= \frac{v(C)}{\zeta(n)} < 2.
\end{aligned} \tag{3.10}$$

Furthermore, denoting the set of *primitive points* of Λ by Λ', it follows from (3.7) and (3.9) that

$$\begin{aligned}
\sum_{\mathbf{u} \in \Lambda \setminus \{\mathbf{o}\}} f(\mathbf{u}) &= \sum_{l=1}^{\infty} \sum_{\mathbf{u} \in \Lambda'} f(l\mathbf{u}) \\
&= \sum_{\mathbf{u} \in \Lambda'} \sum_{l=1}^{\infty} \sum_{k=1}^{\infty} \mu(k)\chi(kl\mathbf{u}) \\
&= \sum_{\mathbf{u} \in \Lambda'} \sum_{m=1}^{\infty} \chi(m\mathbf{u}) \sum_{k|m} \mu(k) \\
&= \sum_{\mathbf{u} \in \Lambda'} \chi(\mathbf{u}).
\end{aligned} \tag{3.11}$$

Then, by Lemma 3.2, (3.10), and (3.11) there is a suitable lattice Λ with determinant 1 such that

$$\sum_{\mathbf{u} \in \Lambda'} \chi(\mathbf{u}) = \sum_{\mathbf{u} \in \Lambda \setminus \{\mathbf{o}\}} f(\mathbf{u}) < \int_{E^n} f(\mathbf{x})d\mathbf{x} < 2.$$

This implies (3.6), and hence Theorem 3.1 is proved. $\square$

Remark 3.1. *The Minkowski-Hlawka theorem is one of the most important results in the geometry of numbers. In the past five decades, different proofs and improvements were achieved by Bateman [1], Cassels [1], Davenport and Rogers [1], Lekkerkerker [1], Macbeath and Rogers [1], [2], [3], Mahler*

[1], *Malyšev* [1], *Rogers* [6], [7], [9], *Sanov* [1], *Schmidt* [1], [4], *Schneider* [1], *Siegel* [1], *Weil* [1], *and others. However, none were able to improve the asymptotic order of the lower bound.*

3.2. Siegel's Mean Value Formula

A positive definite quadratic form $Q(\mathbf{x}) = \mathbf{x}S\mathbf{x}'$ is called *properly reduced* if $s_{ij} \geq 0$ for all index pairs $j = i + 1$ and if

$$Q(\mathbf{z}) \geq s_{ii}$$

whenever $(z_i, z_{i+1}, \ldots, z_n) = 1$. As in the L-S-M reduced case, it can be proved that every positive definite quadratic form is equivalent to a properly reduced one.

Let $\mathcal{A}$ be the family of $n \times n$ matrices A with determinant 1, and let $\mathcal{A}_1$ be the subfamily such that $Q(\mathbf{x}) = \mathbf{x}AA'\mathbf{x}'$ is properly reduced and the trace of A is nonnegative. Then, denote by $\mathcal{P}$ and $\mathcal{P}_1$ the sets of points

$$(a_{11}, a_{12}, \ldots, a_{1n}, a_{21}, a_{22}, \ldots, a_{2n}, \ldots, a_{nn})$$

in E^{n^2} such that $A \in \mathcal{A}$ and $A \in \mathcal{A}_1$, respectively. It follows from the definition of the properly reduced form that $A \in \text{rint}(\mathcal{P}_1)$ if and only if the corresponding form

$$Q(\mathbf{x}) = \mathbf{x}AA'\mathbf{x}' = \mathbf{x}S\mathbf{x}'$$

satisfies both $s_{ij} > 0$ for $j = i + 1$ and $Q(\mathbf{z}) > s_{ii}$ for $\mathbf{z} \in Z_n$ such that $(z_i, z_{i+1}, \ldots, z_n) = 1$ and $(z_i, z_{i+1}, \ldots, z_n) \neq (\pm 1, 0, \ldots, 0)$. Thus, routine argument yields that $\mathcal{P}_1$ is a *fundamental domain* of $\mathcal{P}$. In other words, let $\mathcal{P}_1(U)$ be the subset of $\mathcal{P}$ corresponding to $\{AU : A \in \mathcal{A}_1\}$ and let $\mathcal{U}$ be the family of all $n \times n$ unimodular matrices U with determinant 1. Then

$$\bigcup_{U \in \mathcal{U}} \mathcal{P}_1(U)$$

is a tiling in $\mathcal{P}$. For any *Jordan measurable subset* $\mathcal{J}$ of $\mathcal{P}$, let $\overline{\mathcal{J}}$ be the cone in E^{n^2} with base $\mathcal{J}$ and vertex $\mathbf{o}$. Then, in $\mathcal{P}$ we define a measure σ by

$$\sigma(\mathcal{J}) = v(\overline{\mathcal{J}}).$$

Let $f(\mathbf{x})$ be a bounded Riemann integrable function vanishing outside a bounded region and, for convenience, write

$$F = \int_{E^n} f(\mathbf{x})d\mathbf{x}.$$

Also, for any matrix A we write

$$\Phi(A) = \sum_{\mathbf{z} \in Z_n \setminus \{o\}} f(\mathbf{z}A').$$

With these definitions and notations, in 1945 C.L. Siegel proved the following mean value formula.

Lemma 3.3 (Siegel [1]).

$$\sigma(\mathcal{P}_1) = \frac{1}{n} \prod_{k=2}^{n} \zeta(k)$$

and

$$\int_{\mathcal{P}_1} \Phi(A)d\sigma = \sigma(\mathcal{P}_1)F.$$

Proof. Let I be the $(n-1)$-dimensional cube $\{\mathbf{y} = (y_2, y_3, \ldots, y_n) : 0 \le y_i < 1\}$, and let X be a bounded *Lebesgue measurable set* in E^n such that $v(X) = F$. For convenience, we denote the analogues of $\mathcal{A}$, $\mathcal{A}_1$, $\mathcal{P}$, $\mathcal{P}_1$, and σ for the $(n-1) \times (n-1)$ matrices by $\mathcal{A}^*$, $\mathcal{A}_1^*$, $\mathcal{P}^*$, $\mathcal{P}_1^*$, and σ^*, respectively. In addition, we assume that $x_1 \neq 0$ for any point $\mathbf{x} \in X$.

Let $\mathcal{N}$ be the family of the matrices

$$A = \begin{pmatrix} x_1 & 0 & \cdots & 0 \\ x_2 & & & \\ \vdots & & D & \\ x_n & & & \end{pmatrix} \begin{pmatrix} 1 & y_2 & \cdots & y_n \\ 0 & & & \\ \vdots & & A^* & \\ 0 & & & \end{pmatrix},$$

where $\mathbf{x} \in X$, $\mathbf{y} \in I$, $A^* \in \mathcal{A}_1^*$, and D is a diagonal matrix with diagonal elements $|x_1|^{-1/(n-1)}, \ldots, |x_1|^{-1/(n-1)}, |x_1|^{-1/(n-1)}\text{sign } x_1$, and let $\mathcal{G}$ be its corresponding set of points in E^{n^2}. Clearly, $\mathcal{G}$ is a subset of $\mathcal{P}$.

Let $\chi(\mathbf{x})$ be the characteristic function of X and let $\mathcal{X}(A)$ be the characteristic function of $\mathcal{G}$. Then,

$$\sigma(\mathcal{G}) = \int_{\mathcal{P}} \mathcal{X}(A)d\sigma.$$

By routine computation it follows that

$$\sigma(\mathcal{G}) = \frac{n-1}{n}\sigma^*(\mathcal{P}_1^*) \int_{E^n} \chi(\mathbf{x})d\mathbf{x}$$

$$= \frac{n-1}{n}\sigma^*(\mathcal{P}_1^*)F. \tag{3.12}$$

Next, let $\mathbf{z}$ be a primitive point of Z_n and let $\mathcal{U}(\mathbf{z})$ be the set of all unimodular matrices with determinant 1 having first column $\mathbf{z}'$. For any matrix

$$A \in \bigcup_{U \in \mathcal{U}} \text{rint}(\mathcal{P}_1(U)),$$

by the definitions of I and $\mathcal{P}_1^*$ and routine argument it follows that if $zA' \in X$, then $AU \in \mathcal{N}$ for exactly one $U \in \mathcal{U}(z)$; if $zA' \notin X$, then $AU \notin \mathcal{N}$ for any $U \in \mathcal{U}(z)$. Therefore, one has

$$\sum_{U \in \mathcal{U}} \mathcal{X}(AU) = \sum_{z \in Z'} \chi(zA'), \qquad (3.13)$$

where Z' indicates the set of primitive points of Z_n.

Since $\mathcal{P}_1$ is a fundamental domain in $\mathcal{P}$ with respect to $\mathcal{U}$ and σ is invariant under the transformation $A \mapsto AU$ for any fixed $U \in \mathcal{U}$, we have

$$\sigma(\mathcal{G}) = \int_{\mathcal{P}} \mathcal{X}(A)d\sigma = \int_{\mathcal{P}_1} \sum_{U \in \mathcal{U}} \mathcal{X}(AU)d\sigma.$$

Then, subsitituting (3.12) and (3.13) in this formula, we obtain

$$\int_{\mathcal{P}_1} \sum_{z \in Z'} \chi(zA')d\sigma = \frac{n-1}{n}\sigma^*(\mathcal{P}_1^*)F.$$

Since $\chi(k\mathbf{x})$ is the characteristic function of $\frac{1}{k}X$, we have

$$\int_{\mathcal{P}_1} \sum_{z \in Z_n \setminus \{o\}} \chi(zA')d\sigma = \sum_{k=1}^{\infty} \int_{\mathcal{P}_1} \sum_{z \in Z'} \chi(kzA')d\sigma$$

$$= \frac{n-1}{n}\zeta(n)\sigma^*(\mathcal{P}_1^*)F \qquad (3.14)$$

and therefore

$$\int_{\mathcal{P}_1} \epsilon^n \sum_{z \in Z_n \setminus \{o\}} \chi(\epsilon z A')d\sigma = \frac{n-1}{n}\zeta(n)\sigma^*(\mathcal{P}_1^*)F \qquad (3.15)$$

for every $\epsilon > 0$.

For any lattice Λ with determinant 1 and any point $\mathbf{x}$,

$$\text{card}\,\{\Lambda \cap (I_n + \mathbf{x})\} \leq \text{card}\,\{\Lambda \cap 2I_n\}.$$

Hence, by approximation it follows that

$$\sum_{z \in Z_n \setminus \{o\}} \chi(\epsilon z A') \leq c\,\epsilon^{-n} \sum_{z \in Z_n \setminus \{o\}} \chi'(zA'),$$

where c is a constant depending only on X and n, and $\chi'(\mathbf{x})$ is the characteristic function of the unit cube I_n.

Clearly,

$$\lim_{\epsilon \to 0} \epsilon^n \sum_{z \in Z_n \setminus \{o\}} \chi(\epsilon z A') = v(X) = F.$$

In addition, $c \sum \chi'(\mathbf{z}A')$ is integrable over $\mathcal{P}_1$, on account of (3.14) with χ replaced by χ'. Hence, letting $\epsilon \to 0$ in (3.15) and applying *Lebesgue's dominated convergence theorem*, we have

$$\sigma(\mathcal{P}_1)F = \frac{n-1}{n}\zeta(n)\sigma^*(\mathcal{P}_1^*)F \qquad (3.16)$$

and therefore

$$\sigma(\mathcal{P}_1) = \frac{1}{n}\prod_{k=2}^{n}\zeta(k),$$

which proves the first assertion of our lemma. In addition, it follows from (3.14) and (3.16) that

$$\int_{\mathcal{P}_1}\sum_{\mathbf{z}\in Z_n\setminus\{\mathbf{o}\}}\chi(\mathbf{z}A')d\sigma = \sigma(\mathcal{P}_1)F.$$

By a suitable approximation process, one finds that this formula remains true if $\chi(\mathbf{x})$ is replaced by an arbitrary integrable function $f(\mathbf{x})$, which proves the second assertion of our lemma. Lemma 3.3 is proved. $\square$

Remark 3.2. *Siegel's mean value formula is an improvement of Lemma 3.2. So, one can deduce the Minkowski-Hlawka theorem from it.*

Remark 3.3. *Let $f(\mathbf{x}_1, \mathbf{x}_2, \ldots, \mathbf{x}_k)$ be a nonnegative Borel measurable function defined in the space E^{kn} that vanishes outside a bounded region, and for any $n \times n$ matrix A write*

$$\Psi(A) = \sum_{\mathbf{z}_i \in Z_n} f(\mathbf{z}_1 A, \mathbf{z}_2 A, \ldots, \mathbf{z}_k A).$$

By studying the integral

$$\int_{\mathcal{P}_1}\Psi(A)d\sigma,$$

Rogers and Schmidt have successively improved the Minkowski-Hlawka theorem to

$$\delta^*(C) \geq \frac{n\log 2 + \beta}{2^n},$$

where β is a suitable constant. So far, this is the best known lower bound of this type.

Remark 3.4. *For the most interesting case $C = S_n$, the best known lower bound*

$$\delta^*(S_n) \geq \frac{(n-1)\zeta(n)}{2^{n-1}}$$

is due to Ball [2].

3.3. Sphere Coverings and the Coxeter-Few-Rogers Lower Bound for $\delta(S_n)$

Let Y be a discrete set of points in E^n. We will call $S_n + Y$ a *sphere covering* of E^n if

$$E^n = \bigcup_{y \in Y} (S_n + \mathbf{y}).$$

In addition, if Y is a lattice, it will be called a *lattice sphere covering* of E^n. Let $\mathcal{Y}$ be the family of sets Y such that $S_n + Y$ is a covering of E^n, and let $\mathcal{L}$ be the family of lattices Λ such that $S_n + \Lambda$ is a lattice covering of E^n. Then we define

$$\theta(S_n) = \liminf_{\substack{l \to \infty \\ Y \in \mathcal{Y}}} \frac{\operatorname{card}\{lI_n \cap Y\}\omega_n}{l^n}$$

and

$$\theta^*(S_n) = \liminf_{\substack{l \to \infty \\ \Lambda \in \mathcal{L}}} \frac{\operatorname{card}\{lI_n \cap \Lambda\}\omega_n}{l^n}.$$

Usually, $\theta(S_n)$ and $\theta^*(S_n)$ are called the *density* of the thinnest sphere coverings of E^n and the density of the thinnest lattice sphere coverings of E^n, respectively. It is easy to see that as with the lattice packing density,

$$\theta^*(S_n) = \liminf_{\Lambda \in \mathcal{L}} \frac{\omega_n}{\det(\Lambda)}.$$

Let $S_n + X$ be a sphere packing in E^n with the maximal density $\delta(S_n)$. Without loss of generality, we assume that for any point $\mathbf{y} \in E^n$,

$$\operatorname{int}(S_n + \mathbf{y}) \cap (S_n + X) \neq \emptyset.$$

Then, $2S_n + X$ is a sphere covering in E^n. By the definitions of $\delta(S_n)$ and $\theta(S_n)$, one has

$$2^n \delta(S_n) = \limsup_{l \to \infty} \frac{\operatorname{card}\{lI_n \cap X\}v(2S_n)}{l^n}$$
$$\geq \theta(2S_n) = \theta(S_n)$$

and therefore

$$\delta(S_n) \geq \frac{\theta(S_n)}{2^n}. \tag{3.17}$$

Besides the interest in $\theta(S_n)$ itself, by (3.17) a lower bound for $\theta(S_n)$ will imply a lower bound for $\delta(S_n)$. So, in this section we introduce both an upper bound and a lower bound for $\theta(S_n)$.

Theorem 3.2 (Rogers [8]). *When $n \geq 3$,*

$$\theta(S_n) < n \log n + n \log \log n + 5n.$$

Proof. For convenience, in this proof let S_n be the n-dimensional sphere of volume 1. Let l be a large positive number and let m be a large positive integer. Also, write

$$I = \{(x_1, x_2, \ldots, x_n) : 0 \leq x_i < l\}$$

and $\Lambda = lZ_n$. It is easy to see that $S_n + \Lambda$ is a packing. Let $X = \{x_1, x_2, \ldots, x_m\}$ be a subset of I. We will study the system

$$S_n + X + \Lambda. \tag{3.18}$$

Let $\chi(\mathbf{x})$ be the characteristic function of S_n. Then the characteristic function of the set $E' = E^n \setminus (S_n + X + \Lambda)$ is

$$\chi'(\mathbf{x}) = \prod_{i=1}^{m} \left(1 - \sum_{u \in \Lambda} \chi(\mathbf{x} - \mathbf{x}_i - \mathbf{u}) \right).$$

Since this function is periodic in each coordinate with period l, E' is a set with asymptotic density equal to $v(I \cap E')/v(I)$, where

$$v(I \cap E') = \int_I \chi'(\mathbf{x})d\mathbf{x}.$$

Let M be the mean value of $v(I \cap E')$, taken over all distributions of $\mathbf{x}_1, \mathbf{x}_2, \ldots, \mathbf{x}_m$ throughout the cube I. Then

$$M = \frac{1}{l^{mn}} \int_I \int_I \cdots \int_I \left(\int_I \chi'(\mathbf{x}) \, d\mathbf{x} \right) d\mathbf{x}_1 d\mathbf{x}_2 \cdots d\mathbf{x}_m$$

$$= \frac{1}{l^{mn}} \int_I \left[\prod_{i=1}^{m} \int_I \left(1 - \sum_{u \in \Lambda} \chi(\mathbf{x} - \mathbf{x}_i - \mathbf{u}) \right) d\mathbf{x}_i \right] d\mathbf{x}$$

$$= \frac{1}{l^{mn}} \int_I \left[\prod_{i=1}^{m} \left(l^n - \sum_{u \in \Lambda} \int_{I - \mathbf{x} + \mathbf{u}} \chi(-\mathbf{y})d\mathbf{y} \right) \right] d\mathbf{x}$$

$$= \frac{1}{l^{mn}} l^n \left(l^n - \int_{E^n} \chi(-\mathbf{y})d\mathbf{y} \right)^m$$

$$= l^n \left(1 - \frac{1}{l^n} \right)^m.$$

Hence, there is a set X such that the asymptotic density of the corresponding E' does not exceed $(1 - l^{-n})^m$.

Let η be any number satisfying $0 < \eta \leq 1$, and let m' be the maximal cardinality of the set $Y = \{y_1, y_2, \ldots, y_{m'}\}$ such that

$$\eta S_n + Y + \Lambda \tag{3.19}$$

is a packing in E'. Then, one has

$$m'\eta^n \leq l^n \left(1 - \frac{1}{l^n} \right)^m$$

and therefore

$$m' \le \frac{l^n}{\eta^n}\left(1 - \frac{1}{l^n}\right)^m. \qquad (3.20)$$

Since m' is maximal and $0 < \eta \le 1$, it follows from the choice of the two systems (3.18) and (3.19) that

$$(1 + \eta)S_n + X \cup Y + \Lambda$$

is a covering of E^n. The density of this covering is

$$\frac{(1 + \eta)^n(m + m')}{l^n}.$$

Therefore, by (3.20) and the definition of $\theta(S_n)$,

$$\theta(S_n) \le \min_{l,\,m,\,\eta}\left\{(1 + \eta)^n\left[\frac{m}{l^n} + \frac{1}{\eta^n}\left(1 - \frac{1}{l^n}\right)^m\right]\right\}.$$

Then, taking

$$\frac{m}{l^n} = n\log(1/\eta) \quad\text{and}\quad \eta = \frac{1}{n\log n}$$

successively, one has

$$\begin{aligned}
\theta(S_n) &\le \min_{0<\eta\le 1}\left\{(1 + \eta)^n\left(n\log(1/\eta) + \eta^{-n}e^{-n\log(1/\eta)}\right)\right\}\\
&= \min_{0<\eta\le 1}\left\{(1 + \eta)^n\left(1 + n\log(1/\eta)\right)\right\}\\
&< \min_{0<\eta\le 1}\left\{e^{n\eta}\left(1 + n\log(1/\eta)\right)\right\}\\
&\le e^{1/\log n}\left(1 + n\log(n\log n)\right)\\
&< \left(1 + \frac{2}{\log n}\right)(n\log n + n\log\log n + 1)\\
&< n\log n + n\log\log n + 5n.
\end{aligned}$$

Theorem 3.2 is proved.   □

For the lattice case, we have the following analogue.

Theorem 3.3 (Rogers [11]). *For $n \ge 3$, there is a constant c such that*

$$\theta^*(S_n) \le c\,(n\log_2 n)^{\log_2\sqrt{2\pi e}}.$$

Remark 3.5. *In fact, Rogers' original results dealt with not only spheres, but general convex bodies. Let K be an n-dimensional convex body, $n \ge 3$. Rogers proved that*

$$\theta(K) \le n\log n + n\log\log n + 5n$$

and

$$\theta^*(K) \leq n^{\log_2 n + c \log_2 \log_2 n},$$

where c is a suitable constant.

Let $T = \mathrm{conv}\{\mathbf{v}_0, \mathbf{v}_1, \ldots, \mathbf{v}_n\}$ be a regular simplex with side length $\sqrt{2(n+1)/n}$. The $n+1$ unit spheres $S_n + \mathbf{v}_0$, $S_n + \mathbf{v}_1$, $\ldots$, $S_n + \mathbf{v}_n$ just cover T. We define

$$\tau_n = \frac{\sum_{i=0}^n v((S_n + \mathbf{v}_i) \cap T)}{v(T)}$$

$$= \frac{(n+1)v((S_n + \mathbf{v}_0) \cap T)}{v(T)}.$$

As a counterpart of Theorem 3.2, we have the following result.

Theorem 3.4 (Coxeter, Few, and Rogers [1]).

$$\theta(S_n) \geq \tau_n.$$

To prove this theorem, first let us introduce a basic concept and an important lemma. Let P be a polytope in E^n with vertices $\mathbf{v}_0, \mathbf{v}_1, \ldots, \mathbf{v}_k$, and write

$$V_i = \{\mathbf{v}_i + \lambda(\mathbf{x} - \mathbf{v}_i) : \mathbf{x} \in P, \ \lambda \geq 0\}.$$

Then we call

$$\phi(\mathbf{v}_i) = s(\mathrm{bd}(S_n + \mathbf{v}_i) \cap V_i)$$

the *solid angle* of P at $\mathbf{v}_i$.

Lemma 3.4 (Coxeter, Few, and Rogers [1]).   *Let T be an n-dimensional simplex, with vertices $\mathbf{v}_0, \mathbf{v}_1, \ldots, \mathbf{v}_n$, contained in S_n. Then,*

$$\frac{1}{v(T)} \sum_{i=0}^n \phi(\mathbf{v}_i) \geq n\tau_n,$$

where equality holds when T is a regular simplex inscribed in S_n.

Proof. Let $\mathbf{v}_i = (v_{i1}, v_{i2}, \ldots, v_{in})$ and suppose temporarily that $\mathbf{v}_0 = \mathbf{o}$. We introduce new coordinates $(y_1, y_2, \ldots, y_n)$ related to the original coordinates $(x_1, x_2, \ldots, x_n)$ by the equations

$$x_i = v_{1i}y_1 + v_{2i}y_2 + \cdots + v_{ni}y_n, \quad i = 1, 2, \ldots, n.$$

Then the new coordinates of $\mathbf{v}_0, \mathbf{v}_1, \ldots, \mathbf{v}_n$ are $(0, 0, \ldots, 0)$, $(1, 0, \ldots, 0)$, $(0, 0, \ldots, 1)$, respectively, and

$$v(T) = \frac{\nu}{n!}, \tag{3.21}$$

where ν is the absolute value of the determinant of (v_{ij}). Since

$$\int_{E^n} e^{-\langle \mathbf{x}, \mathbf{x} \rangle} d\mathbf{x} = \frac{n\omega_n \Gamma(n/2)}{2},$$

it follows that

$$\begin{aligned}
\frac{\phi(\mathbf{v}_0)}{n\omega_n} &= \int_{y_i \geq 0} e^{-\langle \mathbf{x}, \mathbf{x} \rangle} d\mathbf{x} \bigg/ \int_{E^n} e^{-\langle \mathbf{x}, \mathbf{x} \rangle} d\mathbf{x} \\
&= \frac{2\nu}{n\omega_n \Gamma(n/2)} \int_0^\infty \int_0^\infty \cdots \int_0^\infty e^{-\langle \mathbf{x}, \mathbf{x} \rangle} dy_1 dy_2 \cdots dy_n \\
&= \frac{2\nu}{n\omega_n \Gamma(n/2)} \int_0^\infty \int_0^\infty \cdots \int_0^\infty e^{-Q(\mathbf{y}, \mathbf{v}_0)} dy_1 dy_2 \cdots dy_n,
\end{aligned}$$

where

$$Q(\mathbf{y}, \mathbf{v}_0) = \sum_{j,\,k=1}^n \langle \mathbf{v}_j - \mathbf{v}_0, \mathbf{v}_k - \mathbf{v}_0 \rangle y_j y_k.$$

In the general case, keeping (3.21) in mind, we have

$$\frac{\phi(\mathbf{v}_i)}{v(T)} = \frac{2n!}{\Gamma(n/2)} \int_0^\infty \int_0^\infty \cdots \int_0^\infty e^{-Q(\mathbf{y}, \varphi_i)} dy_1 dy_2 \cdots dy_n, \qquad (3.22)$$

where

$$Q(\mathbf{y}, \varphi_i) = \sum_{j,\,k=1}^n \langle \mathbf{v}_{\varphi_i(j)} - \mathbf{v}_{\varphi_i(0)}, \mathbf{v}_{\varphi_i(k)} - \mathbf{v}_{\varphi_i(0)} \rangle y_j y_k$$

and φ_i is any one of the $n!$ permutations of the integers $0, 1, \ldots, n$ with $\varphi_i(0) = i$. Let φ be an arbitrary permutation of the integers $0, 1, \ldots, n$. It follows from (3.22) and the definition of ω_n that

$$\frac{n! \sum \phi(\mathbf{v}_i)}{v(T)} = \frac{2n!}{\Gamma(n/2)} \int_0^\infty \int_0^\infty \cdots \int_0^\infty \sum_\varphi e^{-Q(\mathbf{y}, \varphi)} dy_1 dy_2 \cdots dy_n.$$

Then, by the *arithmetic mean and geometric mean inequality*, we have

$$\frac{\sum \phi(\mathbf{v}_i)}{v(T)} \geq \frac{2(n+1)!}{\Gamma(n/2)} \int_0^\infty \int_0^\infty \cdots \int_0^\infty e^{\frac{1}{(n+1)!} \sum_\varphi Q(\mathbf{y}, \varphi)} dy_1 dy_2 \cdots dy_n, \tag{3.23}$$

where equality holds when T is regular.

Writing

$$s^2 = \frac{1}{n(n+1)} \sum_{j,\,k=0}^n \langle \mathbf{v}_j - \mathbf{v}_k, \mathbf{v}_j - \mathbf{v}_k \rangle, \qquad (3.24)$$

since

$$\langle \mathbf{v}_i - \mathbf{v}_j, \mathbf{v}_i - \mathbf{v}_j \rangle = \langle \mathbf{v}_i - \mathbf{v}_k, \mathbf{v}_i - \mathbf{v}_j \rangle + \langle \mathbf{v}_k - \mathbf{v}_j, \mathbf{v}_i - \mathbf{v}_j \rangle,$$

routine computation yields

$$\frac{1}{(n+1)!} \sum_{\varphi} \langle \mathbf{v}_{\varphi(j)} - \mathbf{v}_{\varphi(0)}, \mathbf{v}_{\varphi(k)} - \mathbf{v}_{\varphi(0)} \rangle = \begin{cases} \frac{1}{2}s^2 & \text{if } j \neq k, \\ s^2 & \text{if } j = k. \end{cases} \quad (3.25)$$

Then, it follows from (3.23) and (3.25), with the substitutions $sy_i = x_i$, that

$$\frac{\sum \phi(\mathbf{v}_i)}{v(T)} \geq \frac{2(n+1)!}{\Gamma(n/2)s^n} \int_0^\infty \int_0^\infty \cdots \int_0^\infty e^{-Q(\mathbf{x})} dx_1 dx_2 \cdots dx_n, \quad (3.26)$$

where

$$Q(\mathbf{x}) = \sum_{j \leq k} x_j x_k,$$

and equality holds when T is regular. Since the integral in (3.26) is a constant, by checking a special simplex, (3.26) yields

$$\frac{\sum \phi(\mathbf{v}_i)}{v(T)} \geq n \left(\frac{2(n+1)}{ns^2} \right)^{n/2} \tau_n. \quad (3.27)$$

Since $T \subseteq S_n$, it follows from (3.24) that

$$s^2 = \frac{2}{n(n+1)} \left((n+1) \sum_{i=0}^n \langle \mathbf{v}_i, \mathbf{v}_i \rangle - \left\langle \sum_{i=0}^n \mathbf{v}_i, \sum_{i=0}^n \mathbf{v}_i \right\rangle \right)$$

$$\leq \frac{2(n+1)}{n}, \quad (3.28)$$

where equality holds if T is a regular simplex inscribed in S_n. By (3.27) and (3.28), Lemma 3.4 is proved. $\qquad \square$

Proof of Theorem 3.4. For any positive number ϵ, let $S_n + X$ be a covering of E^n such that

$$\liminf_{l \to \infty} \frac{\text{card}\{lI_n \cap X\}v(S_n)}{l^n} \leq \theta(S_n) + \epsilon. \quad (3.29)$$

Without loss of generality, we assume that

$$\text{card}\{X \cap \text{bd}(\lambda S_n + \mathbf{y})\} \leq n + 1 \quad (3.30)$$

for any positive number $\lambda \leq 1$ and any point $\mathbf{y} \in E^n$.

Let $D(\mathbf{x})$ be the Dirichlet-Voronoi cell associated with X at $\mathbf{x} \in X$, and let V be the set of vertices of these Dirichlet-Voronoi cells. For every $\mathbf{v} \in V$, by (3.30) there are exactly $n+1$ points $\mathbf{x}_0, \mathbf{x}_1, \ldots, \mathbf{x}_n$ of X such that $\mathbf{v}$ is a vertex of $D(\mathbf{x}_i)$. Denote the simplex with vertices $\mathbf{x}_0, \mathbf{x}_1, \ldots, \mathbf{x}_n$ by $T(\mathbf{v})$. The family of simplices $T(\mathbf{v})$, $\mathbf{v} \in V$, tile the space E^n. These simplices are Delone simplices. Together they form a Delone triangulation of the space E^n.

Clearly, it follows from (3.30) that the diameters of the Delone simplices associated with X are bounded by 2 from above. Let l be a sufficiently large positive number. By (3.29) and Lemma 3.4 it follows that

$$
\begin{aligned}
\theta(S_n) + 2\epsilon &\geq \frac{\operatorname{card}\{lI_n \cap X\}s(S_n)}{nl^n} \\
&\geq \frac{\sum_{T(\mathbf{v})\in lI_n} \sum_{i=0}^n \phi(\mathbf{x}_i)}{nl^n} \\
&\geq \frac{\sum_{T(\mathbf{v})\in lI_n} v(T(\mathbf{v}))\tau_n}{l^n} \\
&\geq \frac{v((l-2)I_n)}{l^n}\tau_n
\end{aligned}
$$

and therefore

$$\theta(S_n) \geq \tau_n.$$

Thereom 3.4 is proved.   □

As a direct consequence of (3.17), Theorem 3.4, and Corollary 7.2, we have the following lower bound for $\delta(S_n)$.

Theorem 3.5 (Coxeter, Few, and Rogers [1]).

$$\delta(S_n) \geq \frac{\tau_n}{2^n} \sim \frac{n}{2^n e\sqrt{e}}.$$

Remark 3.6. *This lower bound is weaker than both Schmidt's lower bound mentioned in Remark 3.3 and Ball's lower bound mentioned in Remark 3.4. However, its proof shows a connection between sphere packings and sphere coverings.*

Remark 3.7. *As a counterpart of sphere packing, sphere covering itself is also a fascinating research subject. Many authors have made contributions to the study of $\theta^*(S_n)$. The known exact results are summarized in the following table.*

n	$\theta^*(S_n)$	Author
2	$\frac{2\pi}{3\sqrt{3}}$	Kershner [1]
3	$\frac{5\sqrt{5}\pi}{24}$	Bambah [1], Barnes [2], and Few [2]
4	$\frac{2\pi^2}{5\sqrt{5}}$	Delone and Ryškov [1]
5	$\frac{245\sqrt{35}\pi^2}{3888\sqrt{3}}$	Ryškov and Baranovskii [1]

As for the exact value of $\theta(S_n)$, our only knowledge is

$$\theta(S_2) = \theta^*(S_2) = \frac{2\pi}{3\sqrt{3}},$$

which was discovered by Kershner [1].

3.4. Edmund Hlawka

Edmund Hlawka was born in 1916 in Bruck, Austria. Influenced by his schoolteachers, he became interested in mathematics and physics. In 1934 he became a student at the University of Vienna. At first he studied both mathematics and physics, but soon devoted his time solely to mathematics. In 1938 he obtained a doctorate under the supervision of W. Wirtinger.

Impressed by his dissertation, C.L. Siegel invited him to be an assistant in Göttingen. However, Hlawka preferred to stay in Vienna. In 1948, at the age of 32, he became a professor of mathematics at the University of Vienna, where he worked for 33 years. In 1981 he moved to the Technical University of Vienna, from which he retired in 1987. He was also a guest of many well-known institutions including the Institute for Advanced Study, the University of Paris at Sorbonne, California Institute of Technology, ETH-Zürich, and the University of Freiburg.

Professor Hlawka has made fundamental contributions to the geometry of numbers. In particular, the Minkowski-Hlawka theorem is a classic result in mathematics. It has been improved and generalized by many masters such as Cassels, Davenport, Mahler, Malyšev, Rogers, Schmidt, Schneider, Siegel, and Weil. However, so far, no result is essentially better than his original, as regards the exponent. He has also made influential contributions to the theory of uniform distributions, for example, in discrepancy theory. In 1990, Springer-Verlag published his *Selecta* which contains his most important mathematical works.

For his distinguished contributions to mathematics, Professor Hlawka has received many honors. He is a member of both the Österreichische Akademie der Wissenschaften and the Deutsche Akademie der Naturforscher "Leopoldina," and is a corresponding member of the Reinische-Westfälische Akademie der Wissenschaften, the Bayerische Akademie der Wissenschaften, and the Bologna Academy of Sciences. He has been confered honorary doctorates by Universities of Vienna, Erlangen, Salzburg, and Graz, and the Technical University of Graz. He has also been the recipient of many prizes. For example, the Erwin-Schrödinger prize of the Österreichische Akademie der Wissenschaften, the Gauss-medal of the Akademie der Wissenschaften

Berlin, and the Dannie-Heinemann prize of the Göttinger Akademie der Wissenschaften.

4. Lower Bounds for the Blocking Numbers and the Kissing Numbers of Spheres

4.1. The Blocking Numbers of S_3 and S_4

The *blocking number* $b(K)$ of a convex body K is the smallest number of nonoverlapping translates $K + \mathbf{x}_i$, all of which touch K at its boundary, that can prevent any other translate $K + \mathbf{x}$ from touching K.

Although it is as natural as the kissing number, the blocking number was introduced by Zong [2] only in 1994. Clearly, the problem of determining the blocking number of a convex body itself is both interesting and challenging. However, so far our knowledge of the blocking number is very limited. For a general convex body K it follows from the definition that

$$b(K) \leq k(K). \tag{4.1}$$

As for $k(K)$, one can easily deduce that

$$b(K) = b(D(K)),$$

where $D(K)$ is the difference body of K. Therefore, since $D(K)$ is centrally symmetric, we can deal solely with the centrally symmetric case when we study the blocking numbers of convex bodies. In this way, by applying (1.28) it can be deduced that

$$b(K) = 4$$

for every two-dimensional convex domain.

Clearly, like the kissing number, spheres are the most interesting case for the blocking numbers. In this section we will determine the blocking numbers of S_3 and S_4.

Theorem 4.1 (Dalla, Larman, Mani-Levitska, and Zong [1]).

$$b(S_3) = 6.$$

Proof. By routine argument it can be verified that the six nonoverlapping unit spheres centered at $\pm(2,0,0)$, $\pm(0,2,0)$, and $\pm(0,0,2)$ prevent any other unit sphere from touching the one centered at **o**. Thus,

$$b(S_3) \leq 6. \tag{4.2}$$

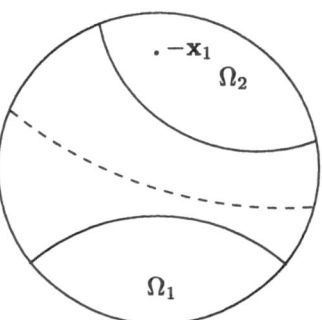

Figure 4.1

On the other hand, suppose that

$$X = \left\{ \mathbf{x}_1, \mathbf{x}_2, \ldots, \mathbf{x}_{b(S_3)} \right\}$$

is a set such that every $S_3 + \mathbf{x}_i$ touches S_3 at its boundary and $S_3 + X$ prevents any other sphere $S_3 + \mathbf{x}$ from touching S_3, and write

$$\Omega_i = \mathrm{bd}(2S_3) \cap (\mathrm{int}(2S_3) + \mathbf{x}_i).$$

Then

$$\mathrm{bd}(2S_3) = \bigcup_{i=1}^{b(S_3)} \Omega_i.$$

Clearly, we have

$$\angle \mathbf{u}_1 \mathbf{o} \mathbf{u}_2 < 2\pi/3$$

whenever both $\mathbf{u}_1$ and $\mathbf{u}_2$ belong to one of the $b(S_3)$ caps Ω_i. Thus, any great circle of $\mathrm{bd}(2S_3)$ intersects at least four of the caps. If, without loss of generality, $-\mathbf{x}_1 \in \Omega_2$, then

$$\Omega_1 \cap \Omega_2 = \emptyset.$$

So, bd($2S_3$) has a great circle that intersects neither Ω_1 nor Ω_2 (see Figure 4.1). Thus, we have

$$b(S_3) \geq 6. \tag{4.3}$$

Our theorem follows from (4.2) and (4.3). $\qquad\qquad\qquad\qquad\qquad\square$

Theorem 4.2 (Dalla, Larman, Mani-Levitska, and Zong [1]).

$$b(S_4) = 9.$$

To prove this theorem we need two general results.

Lemma 4.1. *Let T be a simplex inscribed in S_n. Then $T \cap (rS_n + x)$ has maximal volume for every r if and only if T is regular and $x = o$.*

Proof. Clearly, the lemma is true when $n = 1$. Assume it is true when $n = m - 1$. We proceed to prove it for $n = m$.

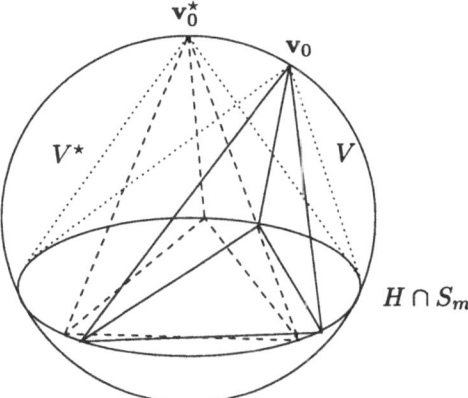

Figure 4.2

Suppose that $T = \text{conv}\{v_0, v_1, \ldots, v_m\}$ is an extreme simplex in E^m but the facet $\text{conv}\{v_1, v_2, \ldots, v_m\}$ is irregular. Let H be the hyperplane determined by $\{v_1, v_2, \ldots, v_m\}$, let $v_0^\star$ be the point of S_m with maximal distance from H, let $\text{conv}\{v_1^\star, v_2^\star, \ldots, v_m^\star\}$ be a regular simplex inscribed in $H \cap S_m$, and write $T^\star = \text{conv}\{v_0^\star, v_1^\star, \ldots, v_m^\star\}$. Also, denote the cones with vertices v_0 and $v_0^\star$ over $H \cap S_m$ by V and $V^\star$, respectively. Clearly, $T \cap (H + tv_0^\star)$ is a simplex inscribed in the $(m-1)$-dimensional sphere $V \cap (H + tv_0^\star)$, $T^\star \cap (H + tv_0^\star)$ is a regular simplex inscribed in the $(m-1)$-dimensional sphere $V^\star \cap (H + tv_0^\star)$ (see Figure 4.2), and the second sphere is not smaller than the first one. Then, we have

$$v(T \cap (rS_m + x))$$

$$= \int_0^\infty s([T \cap (H + tv_0^\star)] \cap [(rS_m + x) \cap (H + tv_0^\star)])dt$$

$$< \int_0^\infty s([T^\star \cap (H + t\mathbf{v}_0^\star)] \cap [(rS_m + \mathbf{x}') \cap (H + t\mathbf{v}_0^\star)])dt$$
$$= v(T^\star \cap (rS_m + \mathbf{x}'))$$

for a suitable r, where $\mathbf{x}'$ is a point on the line determined by $\mathbf{o}$ and $\mathbf{v}_0^\star$ such that $\mathbf{xx}'$ is perpendicular to $\mathbf{v}_0^\star$. Lemma 4.1 follows. $\square$

Lemma 4.2. *Let T' be a spherical simplex lying within a cap Ω with radius ρ of the unit sphere S_n. Then T' has maximal $(n-1)$-dimensional measure if and only if it is regular.*

Proof. Let T be the simplex with the same vertices as T', and let V be the cone with vertex $\mathbf{o}$ over T. The simplex is inscribed in an $(n-1)$-dimensional sphere with center $\mathbf{u}$ and radius ρ. Then, we have

$$s(T') = c \int_V e^{-\|\mathbf{x}\|^2} d\mathbf{x},$$

where

$$c = n\omega_n / \int_{E^n} e^{-\|\mathbf{x}\|^2} d\mathbf{x}.$$

Regarding V as a union of pieces λT, $0 \le \lambda < \infty$, we have

$$s(T') = c \int_0^\infty \left(\int_{\lambda T} e^{-\|\mathbf{x}\|^2} d\lambda T \right) d\lambda$$
$$= c \int_0^\infty \lambda^{n-1} \left(\int_T e^{-\lambda^2 \|\mathbf{x}\|} dT \right) d\lambda. \tag{4.4}$$

In the simplex T, writing $\mathbf{x} = \mathbf{u} + \mathbf{y}$, we have

$$\int_T e^{-\lambda^2 \|\mathbf{x}\|^2} dT = e^{-\lambda^2 \|\mathbf{u}\|} \int_0^\rho e^{-\lambda^2 r^2} \mu(r) dr, \tag{4.5}$$

where $\mu(r)$ indicates the $(n-2)$-dimensional measure of $T \cap (\mathrm{bd}(rS_n) + \mathbf{u})$. Abbreviating $s(T \cap (rS_n + \mathbf{u}))$ to $s(r)$, it is clear that

$$\mu(r) = \frac{ds(r)}{dr}.$$

Then, by Lemma 4.1,

$$\int_0^\rho e^{-\lambda^2 r^2} \mu(r) dr = e^{-\lambda^2 r^2} s(r) \Big|_0^\rho + \int_0^\rho 2\lambda^2 r e^{-\lambda^2 r^2} s(r) dr$$
$$= e^{-\lambda^2 r^2} s(\rho) + \int_0^\rho 2\lambda^2 r e^{-\lambda^2 r^2} s(r) dr$$
$$\le e^{-\lambda^2 r^2} s^\star(\rho) + \int_0^\rho 2\lambda^2 r e^{-\lambda^2 r^2} s^\star(r) dr, \tag{4.6}$$

where s^* indicates the corresponding value in the regular case, and equality holds if and only if T' is regular. Lemma 4.2 follows from (4.4), (4.5), and (4.6). □

Proof of Theorem 4.2. Write $e_1 = (1,0,0,0)$, $e_2 = (0,1,0,0)$, $e_3 = (0,0,1,0)$, $e_4 = (0,0,0,1)$, $e = (1,1,1,1)$,

$$u_i = \frac{2}{\sqrt{1 + 4\epsilon^2 - 2\epsilon}}(e_i - \epsilon e),$$

and

$$v_i = \frac{1}{\sqrt{1 + \epsilon + \epsilon^2}}(-2e_i - \epsilon e).$$

It can be verified that when ϵ is small, $S_4 + \{e, u_1, u_2, u_3, u_4, v_1, v_2, v_3, v_4\}$ blocks any other translate of S_4 from touching it. Thus,

$$b(S_4) \le 9. \tag{4.7}$$

By considering cap coverings, one can deduce $b(S_4) > 8$ from the following statement *"Let P be a four-dimensional polytope with eight facets such that $S_4 \subset P$. Then*

$$\max_{x \in P} \|x\| \ge 2,$$

where equality holds if and only if P is a cube with side length 2." Let P' be the *dual* of P,

$$P' = \{x : \langle x, y \rangle \le 1, \ y \in P\}.$$

This statement is equivalent to the following assertion.

Assertion 4.1. *Let P' be a polytope inscribed in S_4 with eight vertices. Then*

$$\alpha(P') = \min_{x \in \mathrm{bd}(P')} \|x\| \le \tfrac{1}{2},$$

where equality holds if and only if P' is a regular cross polytope I'.

Now we proceed to prove this assertion. It is obvious that $\alpha(I') = \tfrac{1}{2}$ and, for any four-dimensional polytope P such that $S_4 \subset P$,

$$\alpha(P')S_4 \subset P'.$$

As usual, we use f_i to indicate the number of the i-dimensional faces of P'. In addition, without loss of generality, we may assume that P' is *simplicial*. In other words, all its facets are simplices. We study two cases.

Case 1. $\alpha(P') \ge \tfrac{1}{2}$ and $f_3 \le 16$. Projecting the facets of P' from o to the surface of S_4, we get f_3 spherical simplices, say $T'_1, T'_2, \ldots, T'_{f_3}$. Let s^* be the surface area of the corresponding spherical simplex induced by I'. It follows from Lemma 4.2 that

$$s(T'_i) \le s^*,$$

where equality holds if and only if T_i' is induced by a facet of I'. On the other hand, we have

$$\sum_{i=1}^{f_3} s(T_i') = s(S_4).$$

Thus, P' must be a regular cross polytope.

Case 2. $\alpha(P') \geq \frac{1}{2}$ and $f_3 > 16$. By the *Dehn-Sommerville equations* we have

$$\begin{cases} f_0 = -f_0 + 2f_1 - 3f_2 + 4f_3, \\ f_1 = f_1 - 3f_2 + 6f_3, \end{cases}$$

and hence

$$f_1 = f_0 + f_3 \geq 25.$$

Then, since P' has 8 vertices, one vertex of P' is connected to every other by edges. This means that P has a facet F that intersects all the other facets. Let $\mathbf{x}$ be the touching point of F on $\mathrm{bd}(S_4)$, let $p(\mathbf{x})$ be the point of $\mathrm{bd}(P)$ in the opposite direction to $\mathbf{x}$, and assume that

$$\max_{\mathbf{y} \in F} \|\mathbf{o}, \mathbf{y}\| \leq 2.$$

Then simple computation yields that

$$\max_{\mathbf{x} \in P} \|\mathbf{x}\| \geq \|\mathbf{o}, p(\mathbf{x})\| > 2,$$

which contradicts $\alpha(P') \geq \frac{1}{2}$.

As a conclusion, Assertion 4.1 is proved. Thus, we have

$$b(S_4) > 8. \tag{4.8}$$

Then Theorem 4.2 follows from (4.7) and (4.8).   □

To end this section, we propose two conjectures.

Conjecture 4.1. *Let P be an n-dimensional polytope with $2n$ facets and $S_n \subset P$. Then*

$$\max_{\mathbf{x} \in P} \|\mathbf{x}\| \geq \sqrt{n},$$

where equality holds if and only if P is a cube of side length 2.

This conjecture is open for $n \geq 5$.

Conjecture 4.2 (Zong [2]).

$$2n \leq b(K) \leq 2^n.$$

Examples show that both $2n$ and 2^n can be realized by $b(K)$ for certain n-dimensional convex bodies. More surprisingly, they can be realized by convex bodies arbitrarily near to each other, one of which being the cube.

4.2. The Shannon-Wyner Lower Bound for Both $b(S_n)$ and $k(S_n)$

By studying the blocking numbers of spheres, we can deduce the following famous lower bounds.

Theorem 4.3 (Shannon [1] and Wyner [1]).

$$k(S_n) \geq b(S_n) \geq 2^{0.2075\ldots n(1+o(1))}.$$

Proof. Let $\frac{1}{2}S_n + x_i$, $i = 1, 2, \ldots, b(S_n)$, be $b(S_n)$ nonoverlapping spheres, all of which touch $\frac{1}{2}S_n$ at its boundary, and that prevent any other sphere $\frac{1}{2}S_n + x$ from touching $\frac{1}{2}S_n$. Considering the corresponding system S_n, $S_n + x_1, \ldots, S_n + x_{b(S_n)}$, we have

$$\mathrm{bd}(S_n) \subset \bigcup_{i=1}^{b(S_n)} (\mathrm{int}(S_n) + x_i).$$

In other words, all the caps

$$\Omega_i = \mathrm{bd}(S_n) \cap (S_n + x_i)$$

together form a cap covering of $\mathrm{bd}(S_n)$. Thus,

$$s(S_n) \leq \sum_{i=1}^{b(S_n)} s(\Omega_i). \tag{4.9}$$

It is well-known that

$$s(S_n) = n\omega_n = \frac{n\pi^{n/2}}{\Gamma(n/2+1)}. \tag{4.10}$$

On the other hand,

$$
\begin{aligned}
s(\Omega_i) &= \frac{(n-1)\pi^{(n-1)/2}}{\Gamma((n+1)/2)} \int_{1/2}^{1} (1-x^2)^{(n-3)/2} dx \\
&\leq \frac{(n-1)\pi^{(n-1)/2}}{2\Gamma((n+1)/2)} \left(\frac{3}{4}\right)^{(n-3)/2}
\end{aligned}
\tag{4.11}
$$

Then, by (4.9), (4.10), and (4.11), it follows that

$$\frac{n\pi^{n/2}}{\Gamma(n/2+1)} \leq \frac{(n-1)\pi^{(n-1)/2}}{2\Gamma((n+1)/2)} \left(\frac{3}{4}\right)^{(n-3)/2} b(S_n)$$

and therefore

$$b(S_n) \geq \frac{1}{n} \left(\frac{4}{3}\right)^{(n-3)/2} \geq 2^{0.2075\ldots n(1+o(1))}. \tag{4.12}$$

By (4.3) and (4.12), Theorem 4.3 is proved.   □

Remark 4.1. *So far, the Shannon-Wyner lower bound is the best known for $k(S_n)$. However, examining the proof of Theorem 4.3, it is reasonable to conjecture that the exact value of $k(S_n)$ is much larger than $2^{0.2075n}$ when n is large.*

Remark 4.2. *Despite the similarities in their definitions, there is no close relationship between the kissing number and the blocking number except (4.1). Examples of Zong [2] show that when n is sufficiently large, there are two n-dimensional convex bodies K_1 and K_2 in E^n such that*

$$b(K_1) < b(K_2) \quad and \quad k(K_1) > k(K_2).$$

4.3. A Theorem of Swinnerton-Dyer

Despite the regularity of lattices, our knowledge of the asymptotic behavior of $k^*(S_n)$, when n is large, is very limited. In this section we discuss a simple result concerning $k^*(K)$ and the densest lattice packings of a general convex body K in E^n. For this purpose we denote by $k(K, \Lambda)$ the number of translates $K + \mathbf{u}$, $\mathbf{u} \in \Lambda$, that touch K at its boundary.

Theorem 4.4 (Swinnerton-Dyer [1]). *If $K + \Lambda$ is one of the densest lattice packings of K, then*

$$k(K, \Lambda) \geq n(n + 1).$$

Proof. Clearly, $K + \Lambda$ is a packing if and only if $\frac{1}{2}D(K) + \Lambda$ is a packing, and $k(K, \Lambda) = k(\frac{1}{2}D(K), \Lambda)$. Without loss of generality, we may assume that K is centrally symmetric and centered at $\mathbf{o}$. Further we may assume that $\Lambda = Z_n$.

Let $\pm\mathbf{u}_1$, $\pm\mathbf{u}_2$, ..., $\pm\mathbf{u}_m$ be the points of Z_n such that $K + \mathbf{u}_k$, $k = 1$, 2, ..., m, touch K at its boundary. Then,

$$\mathbf{u}_k = (u_{1k}, u_{2k}, \ldots, u_{nk}) \in \mathrm{bd}(2K)$$

for $k = 1$, 2, ..., m. Let I be the $n \times n$ unit matrix. Suppose that $m < n(n+1)/2$. We proceed to show that for suitable choices of an $n \times n$ matrix A and a small number η,

$$\Lambda' = (I + \eta A)Z_n$$

is a packing lattice of K and

$$\det(\Lambda') < 1 = \det(\Lambda).$$

Let $\mathbf{v}_k = (v_{1k}, v_{2k}, \ldots, v_{nk})$ be a unit exterior norm of $2K$ at $\mathbf{u}_k$. Since

$$m + \tfrac{1}{2}n(n+1) < n^2,$$

there is a nonzero symmetric matrix A with elements a_{ij} such that

$$\langle \mathbf{u}_k A, \mathbf{v}_k \rangle = \sum_{i,j} a_{ij} u_{jk} v_{ik} = 0$$

for $k = 1, 2, \ldots, m$. Writing

$$\mathbf{w}_k = \mathbf{u}_k(I + \eta A),$$

it is easy to see that

$$\mathbf{w}_k \notin \operatorname{int}(2K), \quad k = 1, 2, \ldots, m.$$

Thus, $K + \Lambda'$ is a packing of K.

To calculate $\det(\Lambda')$, we have

$$\det(\Lambda') = 1 + c_1 \eta + c_2 \eta^2 + \cdots + c_n \eta^n,$$

where

$$c_1 = \sum_{i=1}^{n} a_{ii}, \quad c_2 = \sum_{i<j} \left(a_{ii} a_{jj} - a_{ij}^2 \right),$$

and

$$2c_2 - c_1^2 = -2 \sum_{i<j} a_{ij}^2 - \sum_{i=1}^{n} a_{ii}^2 < 0.$$

Thus, we can choose a small η with proper sign such that

$$\det(\Lambda') < 1 = \det(\Lambda).$$

By this contradiction Theorem 4.4 is proved.   □

Remark 4.3. *As a counterpart of Swinnerton-Dyer's theorem, Gruber [1] proved that in the sense of Baire category most n-dimensional convex bodies K satisfy*

$$k(K, \Lambda) \leq 2n^2$$

for all of their densest packing lattices. Clearly, by Theorem 2.6 and Remark 2.5 both S_7 and S_8 do not belong to this category.

Remark 4.4. *As a consequence of Swinnerton-Dyer's theorem, we have*

$$k^*(S_n) \geq n(n+1).$$

This lower bound is too small, even when $n = 4$. Inductively, it can be deduced that

$$k^*(S_n) \geq k^*(S_{n-1}) + 2n.$$

For more lower bounds for $k^*(S_n)$ and $k(S_n)$, especially the constructive ones, we refer to the next chapter.

4.4. A Lower Bound for the Translative Kissing Numbers of Superspheres

Let α be a number satisfying $\alpha \geq 1$. The set

$$S_n(\alpha) = \left\{ \mathbf{x} \in E^n : \left(\sum_{i=1}^{n} |x_i|^\alpha \right)^{1/\alpha} \leq 1 \right\}$$

is called an n-dimensional *supersphere*. By *Minkowski's inequality*,

$$\left(\sum_{i=1}^{n} |\theta x_i + (1-\theta) y_i|^\alpha \right)^{1/\alpha} \leq \theta \left(\sum_{i=1}^{n} |x_i|^\alpha \right)^{1/\alpha} + (1-\theta) \left(\sum_{i=1}^{n} |y_i|^\alpha \right)^{1/\alpha}$$

whenever $0 \leq \theta \leq 1$. It follows that $S_n(\alpha)$ is a centrally symmetric convex body. Also, it can be verified that $S_n(1)$ is a cross polytope, $S_n(2)$ is a sphere, and $S_n(\infty)$ is a unit cube. In this section we prove the following result.

Theorem 4.5 (Larman and Zong [1]).

$$3^{0.1072...n(1+o(1))} \leq k(S_n(\alpha)) \leq 3^n.$$

To prove this theorem, first we introduce a basic concept and a simple lemma. Let C be an n-dimensional centrally symmetric convex body centered at $\mathbf{o}$. Then, the Minkowski metric given by C is defined by

$$\|\mathbf{x}, \mathbf{y}\|_C = \begin{cases} 0, & \text{if } \mathbf{x} = \mathbf{y}, \\ \frac{\|\mathbf{x}-\mathbf{y}\|}{\|p(\mathbf{x}-\mathbf{y})\|}, & \text{otherwise}, \end{cases}$$

where $p(\mathbf{x})$ denotes the boundary point of C in the direction of $\mathbf{x}$, and $\|\mathbf{x}\|$ denotes the Euclidean norm of $\mathbf{x}$.

Lemma 4.3 (Zong [5]). *The translative kissing number $k(C)$ of C is the maximal number of points $\mathbf{x}_i \in \mathrm{bd}(C)$ such that*

$$\|\mathbf{x}_i, \mathbf{x}_j\|_C \geq 1, \quad i \neq j.$$

Proof. Suppose $X = \{\mathbf{x}_1, \mathbf{x}_2, \ldots, \mathbf{x}_k\}$ is a subset of $\mathrm{bd}(C)$ such that

$$\|\mathbf{x}_i, \mathbf{x}_j\|_C \geq 1$$

for every pair of distinct points $\mathbf{x}_i$ and $\mathbf{x}_j$. Then, by symmetry it follows that C touches each of the k translates $C + 2\mathbf{x}_i$, $\mathbf{x}_i \in X$, at its boundary, and

$$(\text{int}(C) + 2\mathbf{x}_i) \cap (\text{int}(C) + 2\mathbf{x}_j) = \emptyset$$

whenever $\mathbf{x}_i \neq \mathbf{x}_j$. Therefore,

$$k(C) \geq \text{card}\{X\}.$$

On the other hand, if X is a set of points such that $C + \{\mathbf{o}\} \cup \{2X\}$ is an optimal kissing configuration of C, then $X \subset \text{bd}(C)$, and

$$\|\mathbf{x}, \mathbf{y}\|_C \geq 1$$

holds for every pair of distinct points $\mathbf{x}$ and $\mathbf{y}$ of X. Thus, Lemma 4.3 follows. $\qquad\square$

Proof of Theorem 4.5. It is easy to see that

$$\|\mathbf{x}, \mathbf{y}\|_{S_n(\alpha)} = \left(\sum_{k=1}^{n} |x_k - y_k|^\alpha \right)^{1/\alpha}. \tag{4.13}$$

In order to prove Theorem 4.5, first let us introduce a basic combinatorial result.

Assertion 4.2. *Let m be a positive integer with $m \leq n$. Also let Z^* be a set of integer points $\mathbf{z}_i = (z_{i1}, z_{i2}, \ldots, z_{in})$ with $|z_{ik}| \leq 1$, $\sum_{k=1}^{n} |z_{ik}| = m$, and*

$$\sum_{k=1}^{n} |z_{ik} - z_{jk}| \geq m \tag{4.14}$$

whenever $i \neq j$. Then,

$$k(S_n(\alpha)) \geq \text{card}\{Z^*\}.$$

Taking

$$Y = \left\{ \mathbf{y}_i = m^{-1/\alpha} \mathbf{z}_i : \mathbf{z}_i \in Z^* \right\},$$

we have

$$\left(\sum_{k=1}^{n} |y_{ik}|^\alpha \right)^{1/\alpha} = \left(\frac{1}{m} \sum_{k=1}^{n} |z_{ik}|^\alpha \right)^{1/\alpha}$$

$$= \left(\frac{1}{m} \sum_{k=1}^{n} |z_{ik}| \right)^{1/\alpha} = 1,$$

which means that $\mathbf{y}_i \in \mathrm{bd}(S_n(\alpha))$. On the other hand, by (4.13) and (4.14) it follows that

$$\|\mathbf{y}_i, \mathbf{y}_j\|_{S_n(\alpha)} = \left(\frac{1}{m}\sum_{k=1}^{n}|z_{ik} - z_{jk}|^{\alpha}\right)^{1/\alpha}$$

$$\geq \left(\frac{1}{m}\sum_{k=1}^{n}|z_{ik} - z_{jk}|\right)^{1/\alpha} \geq 1$$

whenever $i \neq j$. Hence, by Lemma 4.3,

$$k(S_n(\alpha)) \geq \mathrm{card}\{Y\} = \mathrm{card}\{Z^*\},$$

which proves Assertion 4.2.

Writing

$$Z^* = \left\{\mathbf{z} \in Z_n : |z_i| \leq 1, \ \sum_{i=1}^{n}|z_i| = m\right\},$$

a simple combinatorial argument yields that

$$\mathrm{card}\{Z^*\} = \binom{n}{m}2^m. \tag{4.15}$$

Similarly, writing $h(m) = \lfloor m/2 \rfloor + 1$, for any point $\mathbf{z} \in Z^*$ there are at most

$$g(n, m) = \binom{m}{h(m)}\binom{n - h(m)}{m - h(m)}2^{m - h(m)} \tag{4.16}$$

points $\mathbf{w} \in Z^*$ such that

$$\sum_{i=1}^{n}|z_i - w_i| < m.$$

Defining

$$f(n, m) = \max\{\mathrm{card}\{Z^*\}\}, \tag{4.17}$$

by (4.15) and (4.16) it follows that

$$\binom{n}{m}2^m - f(n, m)g(n, m) \leq 0$$

and therefore

$$f(n, m) \geq \binom{n}{m}\frac{2^m}{g(n, m)}$$

$$= 2^{h(m)}\binom{n}{m}\binom{m}{h(m)}^{-1}\binom{n - h(m)}{m - h(m)}^{-1}.$$

Writing $l = n/m$, by *Stirling's formula* and detailed computation we have

$$f(n, n/l) \geq \left(2^{1-3/2l}(2l-1)^{1/2l-1}l\right)^{n(1+o(1))}. \tag{4.18}$$

It can be shown by routine computation that the function

$$f(x) = 2^{1-3/2x}(2x-1)^{1/2x-1}x$$

attains its maximum $\frac{9}{8}$ at $x = \frac{9}{2}$. Therefore, by Assertion 4.2, (4.17), and (4.18) it follows that

$$k(S_n(\alpha)) \geq f(n, 2n/9) \geq \left(\frac{9}{8}\right)^{n(1+o(1))}$$
$$= 3^{0.1072\ldots n(1+o(1))}.$$

Theorem 4.5 follows from this together with Theorem 1.1. $\qquad\square$

Remark 4.5. *In* 1985 *Milman* [1] *proved the following theorem: Let $\epsilon > 0$ and let C be an n-dimensional centrally symmetric convex body. Then there are subspaces $E^l \subset E^m$ of E^n with $l \geq c(\epsilon)n$ and an ellipsoid E in E^l such that*

$$E \subseteq p(C \cap E^m) \subseteq (1+\epsilon)E,$$

where $p(\mathbf{x})$ indicates the orthogonal projection from E^n onto E^l. As a corollary to this result, Talata [1] *deduced that there exists a positive constant c such that*

$$k(K) \geq 3^{cn}$$

holds for every n-dimensional convex body K. In fact, Bourgain knew this result much earlier, but he did not publish it.

5. Sphere Packings Constructed from Codes

5.1. Codes

Let F be a *finite field*. A *code* $\aleph$ over F of length n is simply a subset of F^n. A typical point, or *codeword*, of $\aleph$ has the form $\mathbf{u} = (u_1,\ u_2,\ \ldots,\ u_n)$, where $u_i \in F$. For convenience, we say that a codeword or point is of type $[\mu^k|\nu^l|\cdots]$ if $\mu = |u_i|$ for k choices of u_i, $\nu = |u_i|$ for l choices of u_i, etc. The *Hamming distance* between two codewords $\mathbf{u}$ and $\mathbf{v}$ of $\aleph$ is the number of coordinates at which they differ, and is denoted by $\|\mathbf{u}, \mathbf{v}\|_H$. The *weight* of a codeword $\mathbf{u}$, $w(\mathbf{u})$, is the number of its nonzero coordinates. Thus,

$$\|\mathbf{u}, \mathbf{v}\|_H = w(\mathbf{u} - \mathbf{v}).$$

The signal transmission process can be described as

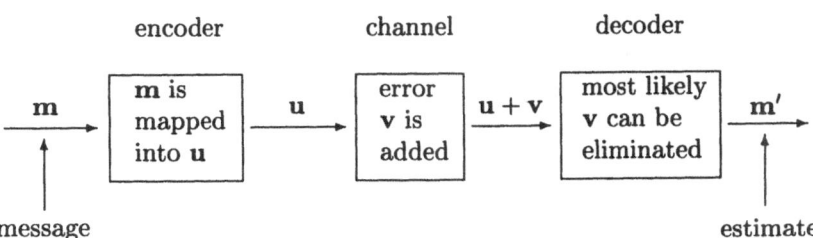

Thus, in this process the number of errors that can be corrected by $\aleph$ is determined by its minimum Hamming distance

$$d(\aleph) = \min_{\substack{\mathbf{u}, \mathbf{v} \in \aleph \\ \mathbf{u} \neq \mathbf{v}}} \|\mathbf{u}, \mathbf{v}\|_H.$$

In coding theory, one of the main problems is to determine $m(F, n, d)$, the maximum number of codewords in any code over F of length n and minimal Hamming distance d. When $F = \{0, 1\}$, determining $m(F, n, d)$ is also an interesting geometric problem. In fact, in this case $m(F, n, d)$ is the maximal number of vertices of the unit cube I_n with the property that any two of them are at Euclidean distance at least $\sqrt{d}$ apart. Usually, a code of length n containing m codewords and with minimal Hamming distance d is said to be an (n, m, d) code. For convenience, we assume that every code contains the zero codeword.

A code $\aleph$ over F is called linear if

$$\mu \mathbf{u} + \nu \mathbf{v} \in \aleph$$

holds whenever $\mathbf{u}$, $\mathbf{v} \in \aleph$ and μ, $\nu \in F$. In other words, a *linear code* is an *F-module*. Then, it is well-known from algebra that $\aleph$ can be spanned by the rows of a *generator matrix*

$$A = (IB),$$

where I is the $k \times k$ unit matrix, and

$$\operatorname{card}\{\aleph\} = q^k,$$

where $q = \operatorname{card}\{F\}$. For convenience, a k-dimensional linear code of length n with minimal Hamming distance d is said to be an $[n, k, d]$ code (or sometimes an $[n, k]$ code).

A code $\aleph$ over F is called cyclic if whenever $(u_0, u_1, \ldots, u_{n-1})$ is a codeword so is $(u_{n-1}, u_0, u_1, \ldots, u_{n-2})$. Unless otherwise stated a *cyclic code* is assumed to be linear. It is convenient to represent a codeword $\mathbf{u} = (u_0, u_1, \ldots, u_{n-1})$ by the polynomial

$$f(x) = u_0 + u_1 x + \cdots + u_{n-1} x^{n-1}$$

in the ring $F_n[x]$ of polynomials modulo $x^n - 1$ with coefficients from F. Then $x f(x)$ represents a cyclic shift of $\mathbf{u}$, and so a linear cyclic code is represented by an ideal in $F_n[x]$. It is well-known from algebra that this ideal can be generated by a single polynomial $g(x)$, a *generator polynomial* for the code. It is easily seen that $g(x)$ divides $x^n - 1$ over F and that the dimension of $\aleph$ is

$$k = n - \deg g(x).$$

To end this section, we introduce several important codes as examples. First of all, we observe a general principle for cyclic codes. Let n be relatively prime to q, and let ξ be a primitive nth root of unity, so that

$$x^n - 1 = \prod_{i=0}^{n-1} \left(x - \xi^i \right).$$

If l is the multiplicative order of q modulo n, then $\xi \in F'$ where F' is a suitable finite field of q^l elements. Also,

$$g(x) = \prod_{j \in J} \left(x - \xi^j \right)$$

for some set $J \subseteq \{0, 1, \ldots, n-1\}$.

Example 5.1 (Binary Hamming Codes). Let $n = 2^l - 1$, $k = n - l$, $d = 3$, and

$$J = \left\{ 1, 2, 2^2, \ldots, 2^{l-1} \right\}.$$

Then the corresponding $g(x)$ generates a binary Hamming code $\mathcal{H}_n$ over $F = \{0, 1\}$. Based on this code, the corresponding extended binary Hamming code $\mathcal{H}_{n+1}$ is defined by

$$\mathcal{H}_{n+1} = \left\{ (u_1, u_2, \ldots, u_{n+1}) : (u_1, u_2, \ldots, u_n) \in \mathcal{H}_n, \sum_{i=1}^{n+1} u_i = 0 \right\}.$$

Example 5.2 (Quadratic Residue Codes). Let p and n be primes such that p is a square modulo n. For $p = 2$, the most important case, this means that n is a prime of the form $8l \pm 1$. Let F be the field $\{0, 1, \ldots, p-1\}$ and let

$$J = \{ j : \ j \neq 0 \text{ is a square modulo } n \}.$$

Then the corresponding $g(x)$ generates a quadratic residue code over F. In particular, the binary *Golay code* $\mathcal{G}_{23}$ is the quadratic residue code over $F = \{0, 1\}$ of length 23. Based on this code, the extended binary Golay code $\mathcal{G}_{24}$ is defined by

$$\mathcal{G}_{24} = \left\{ (u_1, u_2, \ldots, u_{24}) : (u_1, u_2, \ldots, u_{23}) \in \mathcal{G}_{23}, \sum_{i=1}^{24} u_i = 0 \right\}.$$

Example 5.3 (Bose-Chaudhuri-Hocquenghem Codes). The BCH code $\mathcal{B}_{n,d}$ of length n and distance d over F is the cyclic code whose generator polynomial $g(x)$ has roots exactly $\xi, \xi^2, \ldots, \xi^{d-1}$, and their conjugates. Then, routine computation yields that

$$d(\mathcal{B}_{n,d}) \geq d.$$

The special BCH codes over F of length $q - 1$, where $q = \text{card}\{F\}$, are usually called *Reed-Solomon codes*. In these cases,

$$d(\mathcal{B}_{n,d}) = d.$$

Example 5.4 (Justesen Codes). Let $\mathcal{R}$ be a Reed-Solomon code over F, card$\{F\} = 2^l$, with length $n = 2^l - 1$, dimension k, and minimal Hamming distance

$$d(\mathcal{R}) = n - k + 1.$$

Let ν be a primitive element of F. If $\mathbf{u} = (u_0, u_1, \ldots, u_{n-1})$ is a codeword of $\mathcal{R}$, we define $\mathbf{u}^*$ to be the vector

$$\mathbf{u}^* = \left(u_0, u_0, u_1, \nu u_1, \ldots, u_{n-1}, \nu^{n-1} u_{n-1}\right)$$

of length $2n$ over F.

Since F is a vector space of dimension l over $\{0,1\}$, we may set up a one-to-one correspondence between F and $\{0,1\}^l$. Let $\mathbf{u}^\star$ be obtained from $\mathbf{u}^*$ by replacing each component by the corresponding binary l-tuple. Then $\mathbf{u}^\star$ is a binary vector of length $2ln$. The linear code

$$\mathcal{J} = \{\mathbf{u}^\star : \mathbf{u} \in \mathcal{R}\}$$

is called a *Justesen code*. It can be easily shown that its binary dimension is kl.

Example 5.5 (Reed-Muller Codes). Let $m(k)$ be the number of ones in the binary expansion of k. For $1 \le j \le l - 2$, the jth-order binary punctured *Reed-Muller code* $\mathcal{M}_n$ of length $n = 2^l - 1$ is the cyclic code whose generator polynomial has as roots those ξ^k such that

$$1 \le k \le 2^l - 2$$

and

$$1 \le m(k) \le l - j - 1,$$

where ξ is a primitive element of F. Then, we define

$$\mathcal{M}_{n+1} = \left\{(u_1, u_2, \ldots, u_{n+1}) : (u_1, u_2, \ldots, u_n) \in \mathcal{M}_n, \sum_{i=1}^{n+1} u_i = 0\right\}$$

and call it the jth-order Reed-Muller code of length $n + 1 = 2^l$.

5.2. Construction A

Theorem 5.1 (Leech and Sloane [2]). *Let $\aleph$ be an (n, m, d) binary code, and let $h(t)$ be the number of its codewords $\mathbf{u}$ such that $\|\mathbf{u}, \mathbf{o}\|_H = t$. Define*

$$X = \{\mathbf{x} \in E^n : \mathbf{x} \ (\text{mod } 2) \in \aleph\}, \tag{5.1}$$

$$r = \begin{cases} \sqrt{d}/2 & \text{if } d \le 4, \\ 1 & \text{if } d \ge 4, \end{cases}$$

and

$$\tau = \begin{cases} 2^d h(d) & \text{if } d < 4, \\ 2n + 2^4 h(4) & \text{if } d = 4, \\ 2n & \text{if } d > 4. \end{cases}$$

Then $rS_n + X$ is a periodic packing with density

$$\delta(rS_n, X) = \frac{m\omega_n r^\tau}{2^n},$$

in which the sphere rS_n touches τ others.

In particular, X is a lattice if and only if $\aleph$ is a linear code. Suppose $\aleph$ is a linear code of dimension k. Then the corresponding lattice sphere packing has density

$$\delta(rS_n, X) = \frac{\omega_n r^n}{2^{n-k}}.$$

Proof. By its definition, X is a periodic discrete set and can be represented as

$$X = \aleph + 2Z_n, \tag{5.2}$$

where the codewords are regarded as points in E^n.

If $d < 4$, the points closest to the origin o are the $2^d h(d)$ points of type $[1^d | 0^{n-d}]$ obtained from the codewords of weight d. Such points are at Euclidean distance $\sqrt{d}$ from the origin. If $d > 4$, the points closest to the origin are the $2n$ points of type $[2^1 | 0^{n-1}]$, at Euclidean distance 2. Finally, if $d = 4$, both sets of points are at the same distance 2 from the origin. Thus, by (5.2) and routine computation, $rS_n + X$ forms a periodic packing with density

$$\delta(rS_n, X) = \frac{m\omega_n r^n}{2^n}, \tag{5.3}$$

in which rS_n touches τ others.

Since $\aleph$ is a code over $F = \{0, 1\}$, routine argument shows that X is a lattice if and only if $\aleph$ is linear. When $\aleph$ is a linear code of dimension k,

$$m = 2^k.$$

Thus, the last assertion follows from (5.3). Theorem 5.1 is proved. □

The construction defined by (5.1) is usually called *Construction A*.

Example 5.6. Let $\aleph$ be a linear binary code with codewords $\mathbf{u}_0 = (0, 0, 0)$, $\mathbf{u}_1 = (0, 1, 1)$, $\mathbf{u}_2 = (1, 0, 1)$, and $\mathbf{u}_3 = (1, 1, 0)$. Then the lattice given by Construction A is the face-centered cubic lattice that attains both $\delta^*(S_3)$ and $k^*(S_3)$.

Example 5.7. Let $\aleph$ be the extended Hamming code $\mathcal{H}_8$ defined in Example 5.1. By Theorem 5.1 one can construct a lattice sphere packing $S_8 + \Lambda_8$ that attains both $\delta^*(S_8)$ and $k^*(S_8)$.

Example 5.8. In 1954, Golay [1] constructed a nonlinear $(9, 20, 4)$ code. Applying Construction A to this code, Conway and Sloane [1] deduced that

$$k(S_9) \geq 306.$$

Thus, comparing with Theorem 2.6, we have

$$k(S_9) > k^*(S_9).$$

Remark 5.1. *With a* (mod p) *version of Construction A, Rush* [1] *was able to obtain*

$$\delta^*(S_n) \geq 2^{-n(1+o(1))},$$

which is asymptotically the Minkowski-Hlawka lower bound. As shown in Chapter 3, there are several proofs for the Minkowski-Hlawka theorem. However, none can be used to construct lattice sphere packings with such densities, since all the known proofs are based on mean value arguments. Rush's method also shares this drawback, since the codes he has used cannot be clearly constructed.

5.3. Construction B

Theorem 5.2 (Leech and Sloane [2]). *Let $\aleph$ be an (n, m, d) binary code with the property that the weight of each codeword is even, and let $h(t)$ be the number of its codewords $\mathbf{u}$ such that $\|\mathbf{u}, \mathbf{o}\|_H = t$. Define*

$$X = \left\{ \mathbf{x} \in E^n : \mathbf{x} \pmod{2} \in \aleph, \sum_{i=1}^n x_i \in 4Z \right\}, \qquad (5.4)$$

$$r = \begin{cases} \sqrt{d}/2 & \text{if } d \leq 8, \\ \sqrt{2} & \text{if } d \geq 8, \end{cases}$$

and

$$\tau = \begin{cases} 2^{d-1}h(d) & \text{if } d < 8, \\ 2n(n-1) + 2^7 h(8) & \text{if } d = 8, \\ 2n(n-1) & \text{if } d > 8. \end{cases}$$

Then $rS_n + X$ is a periodic packing with density

$$\delta(rS_n, X) = \frac{m\omega_n r^n}{2^{n+1}},$$

and the sphere rS_n touches τ others.

In particular, X is a lattice if and only if $\aleph$ is linear. Suppose $\aleph$ is a linear code of dimension k. Then the corresponding lattice sphere packing has density

$$\delta(rS_n, X) = \frac{\omega_n r^n}{2^{n+1-k}}.$$

This theorem can be proved in a way similar to Theorem 5.1, so its proof is omitted here. Usually, (5.4) is called *Construction B*. With this method one can construct some sphere packings with high densities. Here we give two examples.

Example 5.9. Let $\aleph$ be the first-order Reed-Muller code $\mathcal{M}_{16}$ defined in Example 5.5. By Theorem 5.2 one can construct a lattice Λ_{16} such that $S_{16} + \Lambda_{16}$ is a packing with density

$$\delta(S_{16}, \Lambda_{16}) = \frac{\omega_{16}}{2^4}$$

in which each sphere touches 4320 others. This is the densest lattice sphere packing known in E^{16}.

Example 5.10. Let $\aleph$ be the extended Golay code $\mathcal{G}_{24}$ defined in Example 5.2 and let X be the set constructed by (5.4). Then the set

$$\Lambda_{24} = \tfrac{1}{\sqrt{2}} X \bigcup \left(\mathbf{u} + \tfrac{1}{\sqrt{2}} X \right),$$

where $\mathbf{u} = \frac{1}{\sqrt{8}}(1, 1, \ldots, 1, -3)$, is a lattice known as the *Leech lattice*. This lattice has many important properties. For instance, up to isometry, $S_{24} + \Lambda_{24}$ is the only packing in which every sphere touches

$$k(S_{24}) = k^*(S_{24}) = 196560$$

others (see Theorem 9.4). In addition,

$$\delta(S_{24}, \Lambda_{24}) = \frac{\pi^{12}}{12!},$$

which is the densest lattice sphere packing known in E^{24}.

5.4. Construction C

Let $\mathbf{x} = (x_1, x_2, \ldots, x_n)$ be a point in E^n with integer coordinates. If the component x_i of $\mathbf{x}$ can be expanded as a binary series

$$x_i = \sum_{j=0}^{\infty} x_{ji} 2^j,$$

where complementary notation is used for negative integers, we call the matrix

$$A(\mathbf{x}) = \begin{pmatrix} x_{01} & x_{02} & \cdots & x_{0n} \\ x_{11} & x_{12} & \cdots & x_{1n} \\ x_{21} & x_{22} & \cdots & x_{2n} \\ \vdots & \vdots & \ddots & \vdots \end{pmatrix}$$

the *coordinate array* of $\mathbf{x}$. For example, the coordinate array of $(-2, -1, 1, 2)$ in E^4 is

$$\begin{pmatrix} 0 & 1 & 1 & 0 \\ 1 & 1 & 0 & 1 \\ 1 & 1 & 0 & 0 \\ \vdots & \vdots & \vdots & \vdots \end{pmatrix}.$$

For convenience, letting $\mathbf{u}$ and $\mathbf{v}$ be codewords of a binary code $\aleph$, we say that $\mathbf{v}$ is contained in $\mathbf{u}$ if $u_i = 1$ whenever $v_i = 1$.

Theorem 5.3 (Leech and Sloane [2]). *Suppose $\aleph_0, \aleph_1, \ldots, \aleph_k$ are binary codes of types (n, m_0, d_0), (n, m_1, d_1), $\ldots$, (n, m_k, d_k), respectively, where $d_i = \eta 4^{k-i}$ and $\eta = 1$ or 2. Define*

$$X = \{\mathbf{x} \in Z_n : (x_{j1}, x_{j2}, \ldots, x_{jn}) \in \aleph_j, \ j = 0, 1, \ldots, k\}, \qquad (5.5)$$

$$r = \sqrt{\eta}\, 2^{k-1},$$

and

$$\tau = \sum_{j=0}^{k} \sum_{\mathbf{u} \in \aleph_j} \tau_j(\mathbf{u}),$$

where $\tau_j(\mathbf{u})$ is the number of codewords in $\aleph_{j+1} \cap \aleph_{j+2} \cap \cdots \cap \aleph_k$ that are contained in the codeword $\mathbf{u}$ of $\aleph_j$. Then $rS_n + X$ is a periodic packing with density

$$\delta(rS_n, X) = \frac{\omega_n \eta^{n/2}}{2^{2n}} \prod_{j=0}^{k} m_j$$

in which the sphere rS_n touches τ others.

Proof. Let $\mathbf{x}$ and $\mathbf{y}$ be distinct points in X, and assume that j is the first index such that

$$(x_{j1}, x_{j2}, \ldots, x_{jn}) \neq (y_{j1}, y_{j2}, \ldots, y_{jn}).$$

We now consider two cases.

Case 1. $j > k$. Then the Euclidean distance between $\mathbf{x}$ and $\mathbf{y}$ is at least 2^{k+1}.

Case 2. $0 \leq j \leq k$. Since $\aleph_j$ is an (n, m_j, d_j) code with $d_j = \eta 4^{k-j}$, the two points $\mathbf{x}$ and $\mathbf{y}$ differ in at least d_j coordinates, and therefore the Euclidean distance between them is at least

$$\sqrt{d_j 4^j} = \sqrt{\eta}\, 2^k.$$

So in both cases we have

$$\min_{\substack{\mathbf{x},\mathbf{y} \in X \\ \mathbf{x} \neq \mathbf{y}}} \|\mathbf{x}, \mathbf{y}\| \geq \sqrt{\eta}\, 2^k$$

and $rS_n + X$ is a periodic packing.

As in the proof of Theorem 5.1, it is easy to see that there are

$$m = \prod_{j=0}^{k} m_j$$

points of X in the cube

$$I = \{\mathbf{x} \in E^n : 0 \leq x_i < 2^{k+1}\}.$$

Thus, by the periodic property of X it can be deduced that the packing density of $rS_n + X$ is

$$\delta = \frac{m v(rS_n)}{2^{(k+1)n}} = \frac{\omega_n \eta^{n/2}}{2^{2n}} \prod_{j=0}^{k} m_j.$$

Let X' be the set of points $\mathbf{x} \in X$ such that $rS_n + \mathbf{x}$ touches rS_n at its boundary, and let X_j be the set of points $\mathbf{x} \in X$ of $[(2^j)^{d_j} | 0^{n-d_j}]$ type. For a point $\mathbf{x} \in X$, suppose that j is the first index such that

$$(x_{j1}, x_{j2}, \ldots, x_{jn}) \neq (0, 0, \ldots, 0).$$

Then, by routine argument it follows that both $(x_{j1}, x_{j2}, \ldots, x_{jn})$ and $\mathbf{x}$ are of $[(2^j)^{d_j} | 0^{n-d_j}]$ type if $\mathbf{x} \in X'$. Thus, we have

$$\text{card}\{X'\} = \sum_{j=0}^{k} \text{card}\{X_j\}. \tag{5.6}$$

By the definition of the coordinate array we have

$$x_{li} = \begin{cases} 0 & \text{if } x_j = 2^j, \\ 1 & \text{if } x_j = -2^j \end{cases}$$

for all indices $l \geq j$. Thus, if $\mathbf{x} \in X'$ is a point such that

$$(x_{j1}, x_{j2}, \ldots, x_{jn}) \in \aleph_j,$$

then the minus signs of its coordinates must be at the locations of the ones in some codeword

$$\mathbf{v} \in \aleph_{j+1} \cap \aleph_{j+2} \cap \cdots \cap \aleph_k. \tag{5.7}$$

On the other hand, for any codeword of (5.7) that is contained in a codeword $\mathbf{u} \in \aleph_j$ there is a point $\mathbf{x} \in X$ such that

$$\|\mathbf{x}, \mathbf{o}\| = 2r.$$

Thus,

$$\text{card}\{X_j\} = \sum_{\mathbf{u} \in \aleph_j} \tau_j(\mathbf{u}). \tag{5.8}$$

Then (5.6) and (5.8) together yield

$$\text{card}\{X'\} = \sum_{j=0}^{k} \sum_{\mathbf{u} \in \aleph_j} \tau_j(\mathbf{u}) = \tau,$$

which implies the last assertion of Theorem 5.3. $\square$

Usually, (5.5) is called *Construction C*.

Remark 5.2. *For general codes, the set X constructed by (5.5) is not a lattice. However, if the codes $\aleph_0$, $\aleph_1$, ..., $\aleph_k$ are linear and nested, i.e., $\aleph_j \subseteq \aleph_{j+1}$ for all indices $0 \leq j \leq k-1$, then the set is a lattice.*

Example 5.11. Let $\aleph_j$ be the $2j$th Reed-Muller code of length $n = 2^m$. In addition, take $\eta = 1$ and $k = m/2$ if m is even; and take $\eta = 2$ and $k = (m-1)/2$ if m is odd. Then, by applying Construction C, Leech [2] was able to construct a lattice packing $S_n + \Lambda_n$ in which every sphere touches

$$\tau = \prod_{i=1}^{m} \left(2 + 2^i\right) \sim 4.768 \ldots n^{(m+1)/2}$$

others. So far this is the best known lower bound for $k^*(S_n)$ when $n = 2^m$ is large.

Remark 5.3. *Construction C can be used to obtain dense sphere packings in high dimensions. For example, by applying this construction to certain Justesen codes, Sloane [1] has shown that*

$$\delta(S_n) \geq 2^{-6n(1+o(1))}$$

for integers n of $k2^k$ type. This result is weaker than the Minkowski-Hlawka theorem. Since the Justesen codes used cannot be clearly constructed, Sloane's lower bound is not efficiently constructed.

5.5. Some General Remarks

Besides the previous constructions, there are many other ingenious methods for constructing sphere packings. For example, applying ideas from algebraic curves, Litsyn and Tsfasman [1], [2] obtained a nonlattice packing $S_n + X$ with

$$\delta(S_n, X) \geq 2^{-1.31n(1+o(1))}$$

and a lattice packing $S_n + \Lambda$ with

$$\delta(S_n, \Lambda) \geq 2^{-2.30n(1+o(1))}.$$

For more about sphere packing constructions, we refer to Conway and Sloane [1].

To end this chapter, based upon Example 5.11 we propose the following problems.

Problem 5.1. *Does there exist a positive number c such that*

$$k^*(S_n) \geq 3^{cn}$$

holds for every sufficiently high dimension? If so, what is the largest c?

Very probably the answer to this problem is "no." If so, the following modification would be interesting.

Problem 5.2. *Does there exist a positive number c such that*

$$k^*(S_n) \geq 3^{cn}$$

holds for infinitely many dimensions? If so, what is the largest c?

6. Upper Bounds for the Packing Densities and the Kissing Numbers of Spheres I

6.1. Blichfeldt's Upper Bound for the Packing Densities of Spheres

In 1929 H.F. Blichfeldt discovered a remarkable method for determining an upper bound for $\delta(S_n)$. Roughly speaking, his idea runs as follows. Let $S_n + X$ be a packing and let r be a number such that $r > 1$. Then, replace the spheres $S_n + \mathbf{x}$, where $\mathbf{x} \in X$, by $rS_n + \mathbf{x}$ and fill each of these new spheres with a certain amount of mass, of variable density, such that the total mass at any point of E^n does not exceed 1. Hence, the total mass of the spheres $rS_n + \mathbf{x}$, where $\mathbf{x} \in X \cap (l - r)I_n$, does not exceed the volume of the large cube lI_n. In this way, Blichfeldt obtained the first significant upper bound for $\delta(S_n)$.

Theorem 6.1 (Blichfeldt [1]).

$$\delta(S_n) \leq \frac{n+1}{2} 2^{-0.5n}.$$

In order to prove this theorem, we introduce two fundamental lemmas.

Lemma 6.1. *Let m be a positive integer and let $\mathbf{x}, \mathbf{x}_1, \ldots, \mathbf{x}_m$ be arbitrary points in E^n. Then, one has*

$$\sum_{j,\,k=1}^{m} \|\mathbf{x}_j, \mathbf{x}_k\|^2 \leq 2m \sum_{j=1}^{m} \|\mathbf{x}, \mathbf{x}_j\|^2.$$

Proof. Let a_1, a_2, ..., a_m be m real numbers. Routine analysis yields

$$\sum_{j,\,k=1}^{m} (a_j - a_k)^2 = \sum_{j,\,k=1}^{m} \left(a_j^2 + a_k^2 - 2a_j a_k\right)$$

$$= \sum_{j,\,k=1}^{m} \left(a_j^2 + a_k^2\right) - 2\left(\sum_{j=1}^{m} a_j\right)^2$$

$$\leq \sum_{j,\,k=1}^{m} \left(a_j^2 + a_k^2\right)$$

$$= 2m \sum_{j=1}^{m} a_j^2.$$

Thus, writing $\mathbf{y}_j = \mathbf{x} - \mathbf{x}_j = (y_{j1},\, y_{j2},\, \ldots,\, y_{jn})$, we have

$$\sum_{j,\,k=1}^{m} \|\mathbf{x}_j, \mathbf{x}_k\|^2 = \sum_{j,\,k=1}^{m} \|\mathbf{y}_j, \mathbf{y}_k\|^2$$

$$= \sum_{i=1}^{n} \sum_{j,\,k=1}^{m} (y_{ji} - y_{ki})^2$$

$$\leq 2m \sum_{i=1}^{n} \sum_{j=1}^{m} (y_{ji})^2$$

$$= 2m \sum_{j=1}^{m} \sum_{i=1}^{n} (y_{ji})^2$$

$$= 2m \sum_{j=1}^{m} \|\mathbf{y}_j\|^2$$

$$= 2m \sum_{j=1}^{m} \|\mathbf{x}, \mathbf{x}_j\|^2.$$

Lemma 6.1 is proved.   □

Lemma 6.2 (Blichfeldt [1]). *Let r be a positive number, and let $\sigma(x)$ be a function that is nonnegative and continuous for $0 \leq x \leq r$ and vanishes for $x > r$. Suppose that for each packing $S_n + X$,*

$$\sum_{\mathbf{x}_i \in X} \sigma(\|\mathbf{x}, \mathbf{x}_i\|) \leq 1 \tag{6.1}$$

holds for every point $\mathbf{x} \in E^n$. Then

$$\delta(S_n) \leq J^{-1},$$

where

$$J = n \int_0^r x^{n-1}\sigma(x)dx.$$

Proof. Let l be a large number and suppose that there are exactly m spheres $S_n + \mathbf{x}_1$, $S_n + \mathbf{x}_2$, $\ldots$, $S_n + \mathbf{x}_m$ of the packing $S_n + X$ that are entirely contained in the large cube lI_n. It is obvious that

$$rS_n + \mathbf{x}_i \subset (l + 2r)I_n,$$

and therefore

$$\sigma(\|\mathbf{x}, \mathbf{x}_i\|) = 0 \tag{6.2}$$

holds for every index $i = 1, 2, \ldots, m$ if $\mathbf{x} \notin (l + 2r)I_n$. Then, by (6.1) and (6.2) it follows that

$$v((l + 2r)I_n) \geq \int_{(l+2r)I_n} \sum_{i=1}^{m} \sigma(\|\mathbf{x}, \mathbf{x}_i\|)d\mathbf{x}$$

$$= \sum_{i=1}^{m} \int_{E^n} \sigma(\|\mathbf{x}, \mathbf{x}_i\|)d\mathbf{x}$$

$$= m \int_{E^n} \sigma(\|\mathbf{x}\|)d\mathbf{x}$$

$$= m \int_0^r \sigma(x)d(\omega_n x^n)$$

$$= mn\omega_n \int_0^r x^{n-1}\sigma(x)dx$$

$$= m\omega_n J.$$

Hence, by the definition of $\delta(S_n)$, we have

$$\delta(S_n) = \sup_X \limsup_{r \to \infty} \frac{m\omega_n}{v(lI_n)}$$

$$\leq \sup_X \limsup_{r \to \infty} \frac{v((l + 2r)I_n)}{Jv(lI_n)}$$

$$= \sup_X \limsup_{r \to \infty} \left(1 + \frac{2r}{l}\right)^n J^{-1}$$

$$= J^{-1}.$$

Lemma 6.2 is proved.   □

With these preparations, we proceed to prove Theorem 6.1.

Proof of Theorem 6.1. Let $S_n + X$ be a packing and, for convenience, enumerate the points of X by $\mathbf{x}_1$, $\mathbf{x}_2$, $\ldots$. Since $S_n + X$ is a packing, we have

$$\|\mathbf{x}_j, \mathbf{x}_k\| \geq 2$$

whenever $j \neq k$ and therefore, for each finite set of distinct indices i_1, i_2, $\ldots$, i_m,

$$\sum_{j,\,k=1}^{m} \|\mathbf{x}_{i_j}, \mathbf{x}_{i_k}\|^2 \geq 4m(m-1).$$

Thus, for each point $\mathbf{x} \in E^n$, by Lemma 6.1 we have

$$\sum_{j=1}^{m} \|\mathbf{x}, \mathbf{x}_{i_j}\| \geq 2(m-1). \tag{6.3}$$

Defining

$$\sigma(x) = \max\left\{0, 1 - \tfrac{1}{2}x^2\right\},$$

one has $r = \sqrt{2}$. In addition, for every point $\mathbf{x} \in E^n$, it follows that

$$\begin{aligned}
\sum_{i=1}^{\infty} \sigma(\|\mathbf{x}, \mathbf{x}_i\|) &= \sum_{j=1}^{m} \sigma(\|\mathbf{x}, \mathbf{x}_{i_j}\|) \\
&= \sum_{j=1}^{m} \left(1 - \frac{1}{2}\|\mathbf{x}, \mathbf{x}_{i_j}\|^2\right) \\
&= m - \frac{1}{2}\sum_{j=1}^{m} \|\mathbf{x}, \mathbf{x}_{i_j}\|^2,
\end{aligned}$$

where i_j are the indices i such that $\|\mathbf{x}, \mathbf{x}_i\| < \sqrt{2}$. Hence, by (6.3),

$$\sum_{i=1}^{\infty} \sigma(\|\mathbf{x}, \mathbf{x}_i\|) \leq m - (m-1) = 1.$$

This means that the function $\sigma(x)$ satisfies the condition of Lemma 6.2. Then, routine computation yields that

$$J = n \int_0^{\sqrt{2}} x^{n-1}\left(1 - \frac{1}{2}x^2\right)dx = \frac{2}{n+2}2^{0.5n}.$$

Thus, it follows from Lemma 6.2 that

$$\delta(S_n) \leq J^{-1} = \frac{n+2}{2}2^{-0.5n}.$$

Theorem 6.1 is proved. $\square$

Remark 6.1. *Blichfeldt's method was improved by Rankin* [1], [6], *Sidelnikov* [1], [2], *Levenštein* [1], [2], *and others. In this way, Theorem 6.1 was improved to*

$$\delta(S_n) \leq 2^{-0.5237n(1+o(1))}.$$

By studying the number of spheres that can be packed in an n-dimensional torus, Yudin [1] gave a new proof of this result.

6.2. Rankin's Upper Bound for the Kissing Numbers of Spheres

Let $m(n, \alpha)$ be the maximal number of caps cf *geodesic radius* α that can be packed in $\mathrm{bd}(S_n)$. It is easy to see that

$$k(S_n) = m(n, \pi/6) \tag{6.4}$$

and

$$\delta(S_n) = \lim_{\alpha \to 0} \frac{m(n+1, \alpha)s(\alpha)}{(n+1)\omega_{n+1}}, \tag{6.5}$$

where $s(\alpha)$ indicates the area of a cap of geodesic radius α. Generalizing Blichfeldt's method from E^n to $\mathrm{bd}(S_n)$ and applying (6.4) and (6.5), R.A. Rankin obtained a new proof of Theorem 6.1. In addition, he obtained the following upper bound for $k(S_n)$.

Theorem 6.2 (Rankin [6]).

$$k(S_n) \ll \frac{\sqrt{\pi n^3}}{2} 2^{0.5n},$$

where $f(x) \ll g(x)$ means

$$\limsup_{x \to \infty} \frac{f(x)}{g(x)} \leq 1.$$

First, let us introduce a set of fundamental lemmas.

Lemma 6.3 (Rankin [6]). *Let $\mathbf{x}_j = (x_{j1}, x_{j2}, \ldots, x_{jn})$, $j = 1, 2, \ldots, m$, be m distinct points in $\mathrm{bd}(S_n)$, and define*

$$d = \min_{1 \leq j < k \leq m} \{\|\mathbf{x}_j, \mathbf{x}_k\|\}.$$

Then, for any point $\mathbf{x} \in \mathrm{bd}(S_n)$, we have

$$\left(\sum_{j=1}^m d_j^2\right)^2 - 4m \sum_{j=1}^m d_j^2 + 2m(m-1)d^2 \leq 0,$$

where $d_j = \|\mathbf{x}, \mathbf{x}_j\|$.

Proof. We may suppose, without loss of generality, that $\mathbf{x} = (1, 0, \ldots, 0)$. Then, by the definition of d we have

$$\frac{1}{2}m(m-1)d^2 \le \sum_{i=1}^{n} \sum_{1 \le j < k \le m} (x_{ji} - x_{ki})^2$$

$$= \sum_{i=1}^{n} \left\{ m \sum_{j=1}^{m} (x_{ji})^2 - \left(\sum_{j=1}^{m} x_{ji} \right)^2 \right\}$$

$$= m \sum_{j=1}^{m} (1 - x_{j1})^2 - \left(m - \sum_{j=1}^{m} x_{j1} \right)^2$$

$$+ \sum_{i=2}^{n} \left\{ m \sum_{j=1}^{m} (x_{ji})^2 - \left(\sum_{j=1}^{m} x_{ji} \right)^2 \right\}$$

$$= m \sum_{j=1}^{m} \left\{ (1 - x_{j1})^2 + \sum_{i=2}^{n} (x_{ji})^2 \right\}$$

$$- \left\{ \sum_{j=1}^{m} (1 - x_{j1}) \right\}^2 - \sum_{i=2}^{n} \left(\sum_{j=1}^{m} x_{ji} \right)^2.$$

Since

$$d_j^2 = (1 - x_{j1})^2 + \sum_{i=2}^{n} (x_{ji})^2 = 2(1 - x_{j1}),$$

it follows that

$$\frac{1}{2}m(m-1)d^2 \le m \sum_{j=1}^{m} d_j^2 - \frac{1}{4} \left(\sum_{j=1}^{m} d_j^2 \right)^2,$$

which implies the lemma.   □

Lemma 6.4 (Rankin [6]). *If $0 < \beta < \pi/2$ and $\tan\beta = o(\sqrt{n})$ for large n, then*

$$\int_0^{\beta} (\sin \theta)^{n-2}(\cos \theta - \cos \beta)d\theta \sim \frac{\sec^2 \beta}{n^2}(\sin \beta)^{n+1}.$$

Proof. Substituting

$$t = (n-1) \log \frac{\sin \beta}{\sin \theta},$$

we have

$$\cos \beta \int_0^{\beta} (\sin \theta)^{n-2}d\theta = \frac{(\sin \beta)^{n-1}}{n-1} \int_0^{\infty} \left\{ 1 + \left(1 - e^{\frac{-2t}{n-1}} \right) \tan^2\beta \right\}^{-\frac{1}{2}} e^{-t}dt. \tag{6.6}$$

By *Taylor's theorem* it follows that

$$(1 + x)^{-1/2} = 1 - \frac{1}{2}x + \frac{3}{8}\eta(x)x^2, \tag{6.7}$$

where $0 \leq \eta(x) \leq 1$ for all $x \geq 0$. Thus, taking

$$x = \left(1 - e^{\frac{-2t}{n-1}}\right) \tan^2 \beta$$

and substituting (6.7) in (6.6) we obtain, for a corresponding $\eta(t)$ satisfying $0 \leq \eta(t) \leq 1$,

$$\cos \beta \int_0^\beta (\sin \theta)^{n-2} d\theta = \frac{(\sin \beta)^{n-1}}{n-1} \int_0^\infty e^{-t} \left\{ 1 - \frac{1}{2}\left(1 - e^{\frac{-2t}{n-1}}\right) \tan^2 \beta \right.$$

$$\left. + \frac{3}{8}\eta(t)\left(1 - e^{\frac{-2t}{n-1}}\right)^2 \tan^4 \beta \right\} dt$$

$$= \frac{(\sin \beta)^{n-1}}{n-1}\left\{ 1 - \frac{\tan^2 \beta}{n+1} + \frac{3\eta \tan^4 \beta}{(n+1)(n+3)} \right\},$$

where $0 \leq \eta \leq 1$. Consequently, since $\tan \beta = o(\sqrt{n})$, we have

$$\int_0^\beta (\sin \theta)^{n-2}(\cos \theta - \cos \beta) d\theta = \frac{\sec^2 \beta}{n^2 - 1}(\sin \beta)^{n+1}\left(1 - \frac{3\eta \tan^2 \beta}{n+3}\right)$$

$$\sim \frac{\sec^2 \beta}{n^2}(\sin \beta)^{n+1}.$$

Lemma 6.4 is proved. □

Lemma 6.5 (Rankin [6]). *If $0 < \alpha < \pi/4$ and $\beta = \arcsin(\sqrt{2}\sin\alpha)$, then*

$$m(n, \alpha) \leq \frac{\sqrt{\pi}\,\Gamma((n-1)/2)\sin\beta\tan\beta}{2\,\Gamma(n/2)\int_0^\beta (\sin\theta)^{n-2}(\cos\theta - \cos\beta)d\theta}.$$

Proof. Let α be a positive number with $\alpha < \pi/4$, and suppose that there are $m(n, \alpha)$ nonoverlapping caps $\Omega_i(\alpha)$ of geodesic radius α on $\mathrm{bd}(S_n)$. We replace each cap $\Omega_i(\alpha)$ by a concentric cap $\Omega_i(\beta)$ of geodesic radius β, where

$$\beta = \arcsin(\sqrt{2}\sin\alpha).$$

Also, letting $\gamma = 2\sin\frac{\beta}{2}$ be the Euclidean radius of $\Omega_i(\beta)$, we attach a mass density

$$\sigma_i(r) = \frac{\cos\beta}{\sin^2\beta}(\gamma^2 - r^2) = \frac{2\cos\beta}{\sin^2\beta}(\cos\theta - \cos\beta)$$

with respect to $\Omega_i(\beta)$ to the points $\mathbf{x} \in \mathrm{bd}(S_n)$, where

$$r = \|\mathbf{x}, \mathbf{x}_i\| = 2\sin\frac{\theta}{2}.$$

Then, the total mass of $\Omega_i(\beta)$ is

$$J = \int_0^\beta (n-1)\omega_{n-1}(\sin\theta)^{n-2}\sigma(2\sin\frac{\theta}{2})d\theta$$

$$= 2(n-1)\omega_{n-1}\frac{\cos\beta}{\sin^2\beta}\int_0^\beta (\sin\theta)^{n-2}(\cos\theta - \cos\beta)d\theta. \qquad (6.8)$$

Let $\mathbf{x}$ be a point of bd(S_n). Then either $\mathbf{x}$ belongs to no cap $\Omega_i(\beta)$, in which case the total mass density at $\mathbf{x}$ is zero, or else $\mathbf{x}$ belongs to m such caps centered at $\mathbf{x}_1, \mathbf{x}_2, \ldots, \mathbf{x}_m$, say. In the latter case, the total mass density at $\mathbf{x}$ is, in the notation of Lemma 6.3,

$$\sigma(\mathbf{x}) = \sum_{i=1}^{m} \sigma_i(d_i) = \frac{\cos\beta}{\sin^2\beta}\left(m\gamma^2 - \sum_{i=1}^{m} d_i^2\right).$$

Thus

$$\sum_{i=1}^{m} d_i^2 = 4m\sin^2\frac{\beta}{2} - \sigma(\mathbf{x})\sin\beta\tan\beta$$

$$= 4\left(m - \frac{1+\sec\beta}{2}\sigma(\mathbf{x})\right)\sin^2\frac{\beta}{2},$$

and so, by Lemma 6.3,

$$0 \geq 16\left(m - \frac{1+\sec\beta}{2}\sigma(\mathbf{x})\right)^2 \sin^4\frac{\beta}{2} - 16m\left(m - \frac{1+\sec\beta}{2}\sigma(\mathbf{x})\right)\sin^2\frac{\beta}{2}$$
$$+ 4m(m-1)\sin^2\beta.$$

This reduces to

$$4m(1 - \sigma(\mathbf{x})) \geq \sigma(\mathbf{x})^2\tan^2\beta,$$

which implies $\sigma(\mathbf{x}) \leq 1$.

Therefore, we deduce that $m(n,\alpha)J$ is less than the total surface area of S_n and so, by (6.8),

$$m(n,\alpha) \leq \frac{n\omega_n}{J} = \frac{\sqrt{\pi}\,\Gamma((n-1)/2)\sin\beta\tan\beta}{2\,\Gamma(n/2)\int_0^\beta(\sin\theta)^{n-2}(\cos\theta - \cos\beta)d\theta}.$$

Lemma 6.5 is proved.   □

With these preparations, Theorem 6.2 can be proved as follows.

Proof of Theorem 6.2. It follows from Lemma 6.4 and Lemma 6.5 that

$$m(n;\alpha) \ll \frac{\sqrt{\pi n^3}\sin 2\beta}{2(\sqrt{2}\sin\alpha)^n}.$$

Therefore, by (6.4),

$$k(S_n) = m(n, \pi/6) \ll \frac{\sqrt{\pi n^3}}{2}2^{0.5n}.$$

Theorem 6.2 is proved.   □

Since $\beta = \arcsin(\sqrt{2}\sin\alpha)$, by (6.5), Lemma 6.4, and Lemma 6.5 it follows that

$$
\begin{aligned}
\delta(S_n) &= \lim_{\alpha \to 0} \frac{m(n+1,\alpha) \int_0^\alpha n\omega_n (\sin\theta)^{n-1} d\theta}{(n+1)\omega_{n+1}} \\
&\leq \lim_{\alpha \to 0} \frac{\sin^2\beta \int_0^\alpha (\sin\theta)^{n-1} d\theta}{2 \int_0^\beta (\sin\theta)^{n-1}(\cos\theta - \cos\beta) d\theta} \\
&= \lim_{\alpha \to 0} \frac{(n+1)^2 \int_0^\alpha (\sin\theta)^{n-1} d\theta}{2(\sin\beta)^n} \\
&= \lim_{\alpha \to 0} \frac{(n+1)^2 (\sin\alpha)^n}{2n(\sin\beta)^n} \\
&\sim \frac{n+2}{2} 2^{-0.5n}.
\end{aligned}
$$

Thus, in asymptotic form, we obtain a new proof for Theorem 6.1.

Remark 6.2. *To determine the exact value of $m(n,\alpha)$ for some special n and α is a well-known and very challenging problem. It was proved by Davenport and Hajös [1] that*

$$
m(n,\alpha) = \begin{cases} n+1 & \text{if } \frac{\pi}{4} < \alpha \leq \frac{\pi}{4} + \frac{1}{2}\arcsin\frac{1}{n}, \\ 2n & \text{if } \alpha = \frac{\pi}{4}. \end{cases}
$$

This result implies a remarkable fact: If $n+2$ congruent caps can be packed into $\mathrm{bd}(S_n)$, then so can $2n$.

6.3. An Upper Bound for the Packing Densities of Superspheres

Applying Blichfeldt's method to superspheres $S_n(\alpha)$, we obtain the following theorem.

Theorem 6.3 (van der Corput and Schaake [1], Hlawka [2], Hua [1], and Rankin [2] and [3]). *Writing $\beta = 1/\alpha$,*

$$
\delta(S_n(\alpha)) \leq \begin{cases} \frac{1+\beta n}{2^{\beta n}} & \text{if } \alpha \geq 2, \\ \frac{1+(1-\beta)n}{2^{\beta n}} & \text{if } 1 \leq \alpha \leq 2. \end{cases}
$$

It is easy to see that Theorem 6.3 implies Theorem 6.1, since $S_n = S_n(2)$. In order to prove this theorem, we introduce a generalization of Lemma 6.1.

Lemma 6.6 (Hua [1]). *Let* $a_1, a_2, \ldots, a_m$ *be* m *real numbers. Then*

$$\sum_{i,j=1}^{m} |a_i - a_j|^{\alpha} \leq \begin{cases} 2^{\alpha-1} m \sum_{i=1}^{m} |a_i|^{\alpha} & \text{if } \alpha \geq 2, \\ 2m \sum_{i=1}^{m} |a_i|^{\alpha} & \text{if } 1 \leq \alpha \leq 2. \end{cases}$$

Proof. Defining

$$f(\gamma) = \max_{\Sigma|a_i| \neq 0} \left\{ \left(\sum_{i,j=1}^{m} |a_i - a_j|^{1/\gamma} \right)^{\gamma} \left(m \sum_{i=1}^{m} |a_i|^{1/\gamma} \right)^{-\gamma} \right\}, \qquad (6.9)$$

it follows from a well-known analytical theorem of Riesz [1] (see also Theorem 296 of Hardy, Littlewood, and Pólya [1]) that $\log_2 f(\gamma)$ is a convex function of γ for $0 \leq \gamma \leq 1$. Routine computation yields that

$$f(0) = \max_{\Sigma|a_i| \neq 0} \left\{ \max_{i,j}\{|a_i - a_j|\} / \max_i\{|a_i|\} \right\} = 2,$$

$$f(\tfrac{1}{2}) = \max_{\Sigma|a_i| \neq 0} \left\{ \sqrt{\left[2m \sum_{i=1}^{m} a_i^2 - 2\left(\sum_{i=1}^{m} a_i \right)^2 \right] \left(m \sum_{i=1}^{m} a_i^2 \right)^{-1}} \right\}$$

$$= \sqrt{2},$$

and

$$f(1) \leq \max_{\Sigma|a_i| \neq 0} \left\{ \left(2(m-1) \sum_{i=1}^{m} |a_i| \right) \left(m \sum_{i=1}^{m} |a_i| \right)^{-1} \right\} \leq 2.$$

Therefore, we have

$$\log_2 f(\gamma) \leq \begin{cases} 1 - \gamma & \text{if } 0 \leq \gamma \leq \tfrac{1}{2}, \\ \gamma & \text{if } \tfrac{1}{2} \leq \gamma \leq 1. \end{cases} \qquad (6.10)$$

Substituting $\alpha = 1/\gamma$, it follows from (6.9) and (6.10) that

$$\sum_{i,j=1}^{m} |a_i - a_j|^{\alpha} \leq \begin{cases} 2^{\alpha-1} m \sum_{i=1}^{m} |a_i|^{\alpha} & \text{if } \alpha \geq 2, \\ 2m \sum_{i=1}^{m} |a_i|^{\alpha} & \text{if } 1 \leq \alpha \leq 2. \end{cases}$$

Lemma 6.6 is proved. $\square$

For an arbitrary point $\mathbf{x} = (x_1, x_2 \ldots, x_n)$ of E^n, we define

$$\|\mathbf{x}\|_{\alpha} = \left(\sum_{i=1}^{n} |x_i|^{\alpha} \right)^{1/2}.$$

Meanwhile, we define a density function

$$\sigma(x) = \begin{cases} 1 - \frac{1}{2}x^2 & \text{if } 0 \le x \le \sqrt{2}, \\ 0 & \text{otherwise.} \end{cases}$$

Suppose that $S_n(\alpha) + X$ is a packing, and suppose that there are m points of X, say $\mathbf{x}_1, \mathbf{x}_2, \ldots, \mathbf{x}_m$, such that

$$\|\mathbf{x} - \mathbf{x}_j\|_\alpha < \sqrt{2}, \quad j = 1, 2, \ldots, m.$$

Since

$$\sum_{i=1}^{n} |x_{ji} - x_{ki}|^\alpha \ge 2^\alpha, \quad j \ne k,$$

by Lemma 6.6 and routine calculation it follows that the mass density at $\mathbf{x}$ satisfies

$$\sigma(\mathbf{x}) = \sum_{j=1}^{m} \sigma(\|\mathbf{x} - \mathbf{x}_j\|_\alpha)$$

$$= m - \frac{1}{2} \sum_{j=1}^{m} \sum_{i=1}^{n} |x_i - x_{ji}|^\alpha \le 1.$$

Then, Theorem 6.3 can be proved by an argument that is similar to the proof of Lemma 6.2. We omit the details.

Remark 6.3. *By applying more efficient density functions, the results of Hlawka [2], [3] and Rankin [2], [3] are slightly better than Theorem 6.3.*

6.4. Hans Frederik Blichfeldt

Hans Frederik Blichfeldt was born in Denmark in 1873. In 1888 he took, and passed with high honors, the general preliminary examination conducted by the University of Copenhagen. In the same year, his family emigrated to the United States. During his early American years the young Blichfeldt found employment in Nebraska, Wyoming, Oregon, and Washington, where he worked in the lumber industry. Then for the two years 1892–1894 he worked as a draftsman for the engineering department of the City and County of Whatcom, Washington.

In 1894, with an enthusiastic recommendation from the county superintendent of schools, Blichfeldt was admitted as a special student to Stanford University. There he received a bachelor's degree in mathematics in 1896, followed by a master's degree in 1897. At that time, German universities were important centers of mathematical activity. So, with the help of Rufus Green, Blichfeldt went to Leipzig to work with Sophus Lie. In 1898 he was awarded a doctorate with a dissertation *On a Certain Class of Groups of Transformations in Space of Three Dimensions.*

Blichfeldt returned to Stanford as an instructor in mathematics and remained a member of Stanford's faculty until his retirement in 1938. He was assistant professor of mathematics, 1901–1906; associate professor, 1906–1913; professor, 1913–1938; and professor emeritus from 1938 until his death in 1945. He was also associate professor of mathematics at the University of Chicago, summer quarter 1911, and professor of mathematics at Columbia University for the summer sessions of 1924 and 1925. As the head of the department from 1927 until 1938, he played a very important role in the development of Stanford into a leading mathematical center.

Besides Theorem 2.4 and Theorem 6.1 of this book, Blichfeldt made important contributions to the geometry of numbers and group theory. For example, he solved the problem of finding all finite collineation groups in four variables, a problem whose solution had eluded Klein and Jordan. He published some two dozen research papers of importance. In addition, he was the author or coauthor of three books in group theory.

In 1920, Blichfeldt was elected to the National Academy of Sciences of the United States and was once the vice-president of the American Mathematical Society. For his distinguished scientific contributions, he was awarded a Dannebrogorden in 1938 by the king of Denmark.

7. Upper Bounds for the Packing Densities and the Kissing Numbers of Spheres II

7.1. Rogers' Upper Bound for the Packing Densities of Spheres

Let $T^\star = \text{conv}\{\mathbf{v}_0^\star, \mathbf{v}_1^\star, \ldots, \mathbf{v}_n^\star\}$ be a regular simplex in E^n of side length 2. We define

$$\sigma_n = \frac{\sum_{i=0}^n v(T^\star \cap (S_n + \mathbf{v}_i^\star))}{v(T^\star)}$$
$$= \frac{(n+1)v(T^\star \cap (S_n + \mathbf{v}_0^\star))}{v(T^\star)}.$$

In other words, σ_n is the ratio of the volume of the part of the simplex covered by the unit spheres centered at its vertices to the volume of the whole simplex. Using geometric methods, C.A. Rogers improved Blichfeldt's upper bound in 1958 with the following result.

Theorem 7.1 (Rogers [10]).

$$\delta(S_n) \leq \sigma_n \sim \frac{n}{e} 2^{-0.5n}.$$

This theorem is comparatively easy to imagine, but very difficult to prove, requiring a detailed study of the Dirichlet-Voronoi cells. First, let us recall some basic facts about the Dirichlet-Voronoi cells. Suppose that

$X = \{\mathbf{x}_1, \mathbf{x}_2, \ldots\}$ is a discrete set of points in E^n such that $S_n + X$ is a packing and

$$(S_n + X) \cap \mathrm{int}(S_n + \mathbf{x}) \neq \emptyset \qquad (7.1)$$

for every point $\mathbf{x} \in E^n$. As usual, we denote the Dirichlet-Voronoi cell associated to $\mathbf{x}_j$ by $D(\mathbf{x}_j)$. It is well-known (see Section 1.1) that $D(\mathbf{x}_j)$ is a convex polytope,

$$S_n + \mathbf{x}_j \subset D(\mathbf{x}_j),$$

and the family of Dirichlet-Voronoi cells $D(\mathbf{x}_j)$ together form a tiling of E^n. Also, it follows from (7.1) that

$$d(D(\mathbf{x}_j)) < 4, \quad \mathbf{x}_j \in X.$$

Without loss of generality, we may assume that $\mathbf{x}_j = \mathbf{o}$ and consider the Dirichlet-Voronoi cell $D(\mathbf{o})$. For any sequence

$$F_0 \subset F_1 \subset \cdots \subset F_{n-1}$$

of its faces such that $\dim\{F_i\} = i$ there is a simplex T with vertices $\mathbf{o}$, $\mathbf{v}_1$, $\ldots$, $\mathbf{v}_n$ such that $\mathbf{v}_i \in F_{n-i}$ and

$$\|\mathbf{v}_i, \mathbf{o}\| = \min_{\mathbf{x} \in F_{n-i}} \{\|\mathbf{x}, \mathbf{o}\|\}. \qquad (7.2)$$

If F_{n-i-1}^1, F_{n-i-1}^2, $\ldots$, F_{n-i-1}^k are the $(n-i-1)$-dimensional faces of F_{n-i}, then F_{n-i} is dissected into $\mathrm{conv}\{\mathbf{v}_i, F_{n-i-1}^1\}$, $\mathrm{conv}\{\mathbf{v}_i, F_{n-i-1}^2\}$, $\ldots$, $\mathrm{conv}\{\mathbf{v}_i, F_{n-i-1}^k\}$. Thus, by considering all such sequences of faces, the cell $D(\mathbf{o})$ is dissected into a family of simplices T_1, T_2, $\ldots$, T_l.

Now we introduce a fundamental lemma about the structure of these simplices.

Lemma 7.1. *Let $T = \mathrm{conv}\{\mathbf{o}, \mathbf{v}_1, \ldots, \mathbf{v}_n\}$ be a simplex determined by (7.2). Then*

$$\langle \mathbf{v}_i, \mathbf{v}_j \rangle \geq \frac{2i}{i+1}$$

for $1 \leq i \leq j \leq n$.

Proof. Since $\mathbf{v}_i$ belongs to an $(n-i)$-dimensional face F_{n-i}, which is a subset of at least i facets of $D(\mathbf{o})$, by the definition of the Dirichlet-Voronoi cell there are i points of X, say $\mathbf{x}_1, \mathbf{x}_2, \ldots, \mathbf{x}_i$, such that

$$\langle \mathbf{x}_k - \mathbf{v}_i, \mathbf{x}_k - \mathbf{v}_i \rangle = \langle \mathbf{v}_i, \mathbf{v}_i \rangle, \quad k = 1, 2, \ldots, i. \qquad (7.3)$$

For convenience, we write $\mathbf{x}_0 = \mathbf{o}$. Since $S_n + X$ is a packing and therefore $\|\mathbf{x}_k, \mathbf{x}_l\| \geq 2$ for $k \neq l$, by (7.3) it follows that

$$2i(i+1) = \sum_{0 \leq k \leq l \leq i} 4 \leq \sum_{0 \leq k \leq l \leq i} \langle \mathbf{x}_k - \mathbf{x}_l, \mathbf{x}_k - \mathbf{x}_l \rangle$$

$$= (i+1) \sum_{k=0}^{i} \langle \mathbf{x}_k - \mathbf{v}_i, \mathbf{x}_k - \mathbf{v}_i \rangle$$

$$- \left\langle \sum_{k=0}^{i} (\mathbf{x}_k - \mathbf{v}_i), \sum_{k=0}^{i} (\mathbf{x}_k - \mathbf{v}_i) \right\rangle$$

$$\leq (i+1) \sum_{k=0}^{i} \langle \mathbf{x}_k - \mathbf{v}_i, \mathbf{x}_k - \mathbf{v}_i \rangle$$

$$= (i+1)^2 \langle \mathbf{v}_i, \mathbf{v}_i \rangle.$$

Thus, we have

$$\langle \mathbf{v}_i, \mathbf{v}_i \rangle \geq \frac{2i}{i+1}, \quad i = 1, \ 2, \ \dots, \ n. \tag{7.4}$$

When $1 \leq i < j \leq n$, since $\mathbf{v}_i + \theta(\mathbf{v}_j - \mathbf{v}_i) \in F_{n-i}$ for $0 \leq \theta \leq 1$, it follows from (7.2) that

$$\langle \mathbf{v}_i + \theta(\mathbf{v}_j - \mathbf{v}_i), \mathbf{v}_i + \theta(\mathbf{v}_j - \mathbf{v}_i) \rangle \geq \langle \mathbf{v}_i, \mathbf{v}_i \rangle.$$

Thus, we have

$$2\theta \langle \mathbf{v}_i, \mathbf{v}_j - \mathbf{v}_i \rangle \geq -\theta^2 \langle \mathbf{v}_j - \mathbf{v}_i, \mathbf{v}_j - \mathbf{v}_i \rangle \tag{7.5}$$

for $0 \leq \theta \leq 1$. Since (7.5) holds for any arbitrarily small positive number θ, we deduce that

$$\langle \mathbf{v}_i, \mathbf{v}_j - \mathbf{v}_i \rangle \geq 0. \tag{7.6}$$

Then, it follows from (7.4) and (7.6) that

$$\langle \mathbf{v}_j, \mathbf{v}_i \rangle = \langle \mathbf{v}_i, \mathbf{v}_i \rangle + \langle \mathbf{v}_i, \mathbf{v}_j - \mathbf{v}_i \rangle \geq \langle \mathbf{v}_i, \mathbf{v}_i \rangle \geq \frac{2i}{i+1}.$$

Lemma 7.1 is proved. $\qquad\qquad\qquad\qquad\qquad\qquad\qquad\qquad\qquad$ □

Proof of Theorem 7.1. We suppose that

$$\delta(S_n) > \sigma_n$$

and proceed to deduce a contradiction. By the definition of $\delta(S_n)$ it follows that there is a periodic packing $S_n + X$ such that

$$\delta(S_n, X) > \sigma_n.$$

Then, there is a point $\mathbf{x}_j \in X$ that

$$\frac{v(S_n + \mathbf{x}_j)}{v(D(\mathbf{x}_j))} > \sigma_n. \tag{7.7}$$

For convenience, we assume that $\mathbf{x}_j = \mathbf{o}$. Suppose that $D(\mathbf{o})$ is dissected into a family of simplices $T_1, T_2, \ldots, T_l$ as described in the paragraph preceding Lemma 7.1. Then (7.7) is equivalent to

$$\frac{\sum_{k=1}^{l} v(T_k \cap S_n)}{\sum_{k=1}^{l} v(T_k)} > \sigma_n,$$

which implies

$$\frac{v(T \cap S_n)}{v(T)} > \sigma_n, \tag{7.8}$$

where $T = T_k$ for a certain index k. For convenience, we assume that $T = \mathrm{conv}\{\mathbf{o}, \mathbf{v}_1, \ldots, \mathbf{v}_n\}$, which satisfies (7.2).

Let T^* be the n-dimensional simplex with vertices $(\sqrt{2}, 0, \ldots, 0)$, $(0, \sqrt{2}, 0, \ldots, 0)$, $\ldots$, $(0, 0, \ldots, 0, \sqrt{2})$ in E^{n+1}, and let H be the hyperplane determined by these points. In a similar way to the process of dissecting $D(\mathbf{o})$ into simplices $T_1, T_2, \ldots, T_l$, one can dissect T^* into $(n+1)!$ simplices $T_1^\star, T_2^\star, \ldots, T_{(n+1)!}^\star$, each of which is congruent with $T^* = \mathrm{conv}\{\mathbf{v}_0^*, \mathbf{v}_1^*, \ldots, \mathbf{v}_n^*\}$, where

$$\mathbf{v}_i^* = \left(\sqrt{2}/(i+1), \ldots, \sqrt{2}/(i+1), 0, \ldots, 0\right).$$

It is easy to verify that if $1 \le i \le j \le n$, then

$$\langle \mathbf{v}_i^* - \mathbf{v}_0^*, \mathbf{v}_j^* - \mathbf{v}_0^* \rangle = \frac{2ij + 2i}{(i+1)(j+1)} = \frac{2i}{i+1}.$$

Hence, by Lemma 7.1,

$$\langle \mathbf{v}_i, \mathbf{v}_j \rangle \ge \langle \mathbf{v}_i^* - \mathbf{v}_0^*, \mathbf{v}_j^* - \mathbf{v}_0^* \rangle \tag{7.9}$$

for $1 \le i \le j \le n$.

Let L be the linear transformation from E^n to H determined by

$$\sum_{i=1}^{n} \lambda_i \mathbf{v}_i \mapsto \mathbf{v}_0^* + \sum_{i=1}^{n} \lambda_i (\mathbf{v}_i^* - \mathbf{v}_0^*).$$

Clearly, $L(T) = T^*$. In addition, $L(S_n) = E$ is an ellipsoid in H centered at $\mathbf{v}_0^*$. Now, if $\mathbf{x}$ is a point of $T \cap S_n$, we have

$$\mathbf{x} = \sum_{i=1}^{n} \lambda_i \mathbf{v}_i,$$

where $\lambda_i \ge 0$, and

$$\langle \mathbf{x}, \mathbf{x} \rangle \le 1. \tag{7.10}$$

Hence, by (7.9) and (7.10), the corresponding point $\mathbf{y} = L(\mathbf{x}) \in T^* \cap E$ satisfies

$$
\begin{aligned}
\langle \mathbf{y} - \mathbf{v}_0^*, \mathbf{y} - \mathbf{v}_0^* \rangle &= \sum_{i=1}^{n} \sum_{j=1}^{n} \lambda_i \lambda_j \langle \mathbf{v}_i^* - \mathbf{v}_0^*, \mathbf{v}_j^* - \mathbf{v}_0^* \rangle \\
&\leq \sum_{i=1}^{n} \sum_{j=1}^{n} \lambda_i \lambda_j \langle \mathbf{v}_i, \mathbf{v}_j \rangle \\
&= \langle \mathbf{x}, \mathbf{x} \rangle \leq 1,
\end{aligned}
$$

which implies

$$
T^* \cap E \subseteq T^* \cap (S_n + \mathbf{v}_0^*). \tag{7.11}
$$

Since L preserves the ratio of two volumes, by (7.8) and (7.11) it follows that

$$
\sigma_n < \frac{v(T \cap S_n)}{v(T)} = \frac{v(T^* \cap E)}{v(T^*)} \leq \frac{v(T^* \cap (S_n + \mathbf{v}_0^*))}{v(T^*)}. \tag{7.12}
$$

On the other hand, it follows from the definition of σ_n and the fact that all simplices T_i^* are congruent with T^* that

$$
\frac{v(T^* \cap (S_n + \mathbf{v}_0^*))}{v(T^*)} = c_n,
$$

which contradicts (7.12). This contradiction shows that

$$
\delta(S_n) \leq \sigma_n.
$$

Then, Theorem 7.1 follows from *Daniel's asymptotic formula* (see Corollary 7.1). □

Remark 7.1. *Rogers' upper bound is better than Blichfeldt's only up to a certain constant. However, Rogers' method reveals the geometric nature of the problem.*

7.2. Schläfli's Function

To study the spherical geometry, Schläfli [1] introduced a remarkable function

$$
F_n(\alpha) = \frac{2^n s(n, \alpha)}{n! n \omega_n},
$$

where $s(n, \alpha)$ is the area of a regular *spherical simplex* in $\mathrm{bd}(S_n)$ of *dihedral angle* 2α. This function has been studied by many authors. To give a general impression of this geometric function, we list a few of its well-known properties without proof.

$$F_n(\tfrac{1}{2}\operatorname{arcsec}(n-1)) = 0,$$

$$F_n(\alpha) + F_n(\pi - \alpha) = \frac{2^n}{n!},$$

and

$$F_n(\alpha) = \frac{2}{\pi} \int_{\frac{1}{2}\operatorname{arcsec}(n-1)}^{\alpha} F_{n-2}(\beta)d\beta,$$

where $\sec 2\beta = \sec 2\theta - 2$ and $F_0(\alpha) = F_1(\alpha) = 1$.

In this section we prove the following asymptotic formula.

Lemma 7.2 (Rogers [12]). *Let α be a positive number such that $\alpha < \pi/4$ and $\sec 2\alpha - n + 1$ is bounded. Then, writing $c = \big(\sec 2\alpha - (n-1)\big)^{-1}$,*

$$F_n(\alpha) \sim \frac{\sqrt{1+cn}}{\sqrt{2}\,e^{1/c}n!} \left(\frac{2e}{\pi cn}\right)^{n/2}.$$

Proof. In E^n, take $\mathbf{q} = (q,\,q,\,\ldots,\,q)$, $\mathbf{p}_1 = (p,\,0,\,\ldots,\,0)$, $\mathbf{p}_2 = (0,\,p,\,\ldots,\,0)$, $\ldots$, $\mathbf{p}_n = (0,\,0,\,\ldots,\,0,\,p)$ such that $p > 0$, $q < 0$,

$$\mathbf{p}_i \in \operatorname{bd}(S_n) + \mathbf{q}, \quad i = 1,\,2,\,\ldots,\,n,$$

and the dihedral angle of the regular spherical simplex $T = \mathbf{p}_1\mathbf{p}_2\cdots\mathbf{p}_n$ in $\operatorname{bd}(S_n) + \mathbf{q}$ is 2α. For convenience, we write

$$V = \left\{\mathbf{q} + \sum_{i=1}^{n} \lambda_i(\mathbf{p}_i - \mathbf{q}) : \lambda_i \geq 0\right\}.$$

Since

$$\frac{\int_V e^{-\|\mathbf{x},\mathbf{q}\|^2}d\mathbf{x}}{\int_{E^n} e^{-\|\mathbf{x},\mathbf{q}\|^2}d\mathbf{x}} = \frac{s(n,\alpha)}{n\omega_n}$$

and

$$\int_{E^n} e^{-\|\mathbf{x},\mathbf{q}\|^2}d\mathbf{x} = \frac{n\omega_n\Gamma(n/2)}{2},$$

it follows that

$$F_n(\alpha) = \frac{2^{n+1}}{n!n\omega_n\Gamma(n/2)} \int_V e^{-\|\mathbf{x},\mathbf{q}\|^2}d\mathbf{x}. \tag{7.13}$$

Then, applying a linear transformation L determined by $L(\mathbf{q}) = \mathbf{o}$ and $L(\mathbf{p}_i) = \mathbf{p}_i/p$ for $i = 1,\,2,\,\ldots,\,n$, it follows from (7.13) that

$$F_n(\alpha) = \frac{2^n\sqrt{1+cn}}{\pi^{n/2}n!} \int_0^\infty \cdots \int_0^\infty e^{-Q(\mathbf{y})}dy_1\cdots dy_n, \tag{7.14}$$

where

$$Q(\mathbf{y}) = \sum_{i=1}^{n} y_i^2 + c\left(\sum_{i=1}^{n} y_i\right)^2.$$

When 2α is acute and c is positive, we can apply the well-known result

$$\sqrt{\pi c}\, e^{-cy^2} = \int_{-\infty}^{\infty} e^{-w^2/c + 2wyi}\, dw$$

with $y = \sum_{i=1}^{n} y_i$ to express the integrand of (7.14) in a factorizable form. Then, substituting $w = u + vi$ with $v = \sqrt{cn/2}$, we obtain

$$F_n(\alpha) = \frac{2^n \sqrt{1+cn}}{\pi^{n/2} n! \sqrt{cn}} \int_0^{\infty} \cdots \int_0^{\infty} \left(\int_{-\infty}^{\infty} e^{-Q(\mathbf{y},w)}\, dw\right) dy_1 \cdots dy_n$$

$$= \frac{2^n \sqrt{1+cn}\, e^{n/2 - 1/c}}{\sqrt{\pi c}\, (2\pi cn)^{n/2} n!} \int_{-\infty}^{\infty} e^{(1-u^2)/c - nui/v} \Psi(u,v)^n\, du, \qquad (7.15)$$

where

$$Q(\mathbf{y},w) = \sum_{i=1}^{n} y_i^2 + w^2/c - 2wi \sum_{i=1}^{n} y_i$$

and

$$\Psi(u,v) = 2v \int_0^{\infty} e^{-y^2 - 2(v-ui)y}\, dy.$$

Now we can get an asymptotic expansion for $\Psi(u,v)$ in inverse powers of v using integration by parts. Since the first term of the expansion is 1, we can use the logarithmic series to convert this into an asymptotic expansion for $\log \Psi(u,v)$. The first terms of this expansion take the form

$$\frac{ui}{v} - \frac{u^2 + 1}{cn}.$$

When we substitute the asymptotic expansion for $\log \Psi(u,v)$ in the formula

$$e^{(1-u^2)/c - nui/v} \Psi^n(u,v) = \exp\left(\frac{1-u^2}{c} - \frac{nui}{v} + n\log \Psi(u,v)\right)$$

and apply the exponential expansion, we obtain an expansion for the integrand of (7.15) as a product of $e^{-2u^2/c}$ with an asymptotic expansion in inverse powers of v. Integrating this expansion term by term, we obtain the following formula:

$$F_n(\alpha) \sim \frac{\sqrt{1+cn}}{\sqrt{2}\, e^{1/c} n!} \left(\frac{2e}{\pi cn}\right)^{n/2} \Phi_n(\alpha), \qquad (7.16)$$

where

$$\Phi_n(\alpha) = \sum_{l=0}^{\infty} \frac{1}{l!}\, [n\Pi(\varpi, c)]^l,$$

$$\Pi(\varpi, c) = -\frac{\varpi_1 i}{\sqrt{2cn}} + \frac{\varpi_2 + 4}{4cn}$$

$$+ \sum_{t=1}^{\infty} \frac{(-1)^{t-1}}{t} \left[\sum_{r=1}^{\infty} \left(\frac{1}{2cn} \right)^{r/2} i^r \sum_{0 \leq s \leq r/2} \frac{r!}{s!(r-2s)!} \varpi_{r-2s} \right]^t,$$

and

$$\varpi_m = \begin{cases} 0 & \text{if } m \text{ is odd,} \\ c^{m/2}(m-1)(m-3) \cdots 3 & \text{if } m \text{ is even.} \end{cases}$$

As n tends to infinity, since $1/c$ is bounded, it is easy to see that the asymptotic expansion is in powers of $1/n$, and that it is valid, in the sense that the error made in omitting the terms after that in $1/n^h$ is of order $1/n^{h+1}$. Thus,

$$\Phi_n(\alpha) = 1 + \frac{1}{12}\left(1 + \frac{12}{c} + \frac{18}{c^2} \right) \frac{1}{n} + O\left(\frac{1}{n^2} \right). \tag{7.17}$$

Then, (7.16) and (7.17) together yield Lemma 7.2. $\square$

Remark 7.2. *For an alternative proof of this formula we refer to Böröczky Jr. and Henk* [2].

Corollary 7.1 (Daniel's Asymptotic Formula).

$$\sigma_n \sim \frac{n}{e} 2^{-0.5n}.$$

Proof. Let $T = \text{conv}\{o, v_1, v_2, \ldots, v_n\}$ be a regular simplex in E^n with side length 2, and denote its dihedral angle by 2α. Then,

$$\sigma_n = \frac{(n+1)s(n, \alpha)}{nv(T)} = \frac{(n+1)!n\omega_n F_n(\alpha)}{2^n nv(T)}.$$

It is well-known that

$$v(T) = \frac{2^{n/2}\sqrt{n+1}}{n!}$$

and

$$c = \left(\sec 2\alpha - (n-1) \right)^{-1} = 1.$$

Thus, by Lemma 7.2 and Stirling's formula,

$$\sigma_n \sim \frac{(n+1)!e^{n/2-1}}{\sqrt{2}\,\Gamma(n/2+1)(4n)^{n/2}} \sim \frac{n}{e} 2^{-0.5n}.$$

Corollary 7.1 is proved. $\square$

Corollary 7.2. *Let τ_n be the number defined in Section 3.3. Then*

$$\tau_n \sim \frac{n}{e\sqrt{e}}.$$

Proof. It is easy to see that

$$\tau_n = \left(\frac{2n}{n+1} \right)^{n/2} \sigma_n.$$

Therefore, Corollary 7.2 follows immediately from Corollary 7.1.   □

7.3. The Coxeter-Böröczky Upper Bound for the Kissing Numbers of Spheres

Let $\|\mathbf{x}, \mathbf{y}\|_g$ be the *geodesic metric* defined in $\mathrm{bd}(S_n)$, and let $\widehat{\mathbf{x}\mathbf{y}}$ be the spherical segment from $\mathbf{x}$ to $\mathbf{y}$. For a positive number $\phi < \pi/4$ and a point $\mathbf{x} \in \mathrm{bd}(S_n)$, let $\Omega(\mathbf{x}, \phi)$ be the cap of geodesic radius ϕ and center $\mathbf{x}$, and write

$$\mu(n, \phi) = s(\Omega(\mathbf{x}, \phi)).$$

Let $T = \mathbf{v}_1 \mathbf{v}_2 \ldots \mathbf{v}_n$ be a regular spherical simplex in $\mathrm{bd}(S_n)$ of geodesic side length 2ϕ. In a similar way to the number σ_n, we define

$$\sigma_n(\phi) = \frac{\sum_{i=1}^{n} s(\Omega(\mathbf{v}_i, \phi) \cap T)}{s(T)}$$

$$= \frac{n\, s(\Omega(\mathbf{v}_1, \phi) \cap T)}{s(T)}.$$

Also, let $m(n, \phi)$ be the maximal number of nonoverlapping caps of geodesic radius ϕ that can be packed into $\mathrm{bd}(S_n)$. As an analogue of Theorem 7.1, H.S.M. Coxeter in 1961 made the following conjecture.

Conjecture 7.1 (Coxeter [4]).

$$m(n, \phi) \le \frac{n\omega_n \sigma_n(\phi)}{\mu(n, \phi)} = \frac{2F_{n-1}(\alpha)}{F_n(\alpha)}, \tag{7.18}$$

where $F_{n-1}(\alpha)$ and $F_n(\alpha)$ are Schläfli's functions and

$$\sec 2\alpha = \sec 2\phi + n - 2.$$

This conjecture was proved by Böröczky [1]. As a consequence of (7.18) and Lemma 7.2, since

$$k(S_n) = m(n, \pi/6), \tag{7.19}$$

one can deduce the following upper bound for the kissing numbers of spheres.

Theorem 7.2 (Böröczky [1] and Coxeter [4]).

$$k(S_n) \ll \frac{\sqrt{\pi n^3}}{e\sqrt{2}} 2^{0.5n}.$$

Böröczky's proof of (7.18) is elementary but rather technical.

Let $X = \{\mathbf{x}_1, \mathbf{x}_2, \ldots, \mathbf{x}_m\}$ be a discrete subset of $\mathrm{bd}(S_n)$ such that the caps $\Omega(\mathbf{x}_i, \phi)$, $i = 1, 2, \ldots, m$, together form a packing in $\mathrm{bd}(S_n)$. Analogously to the Euclidean case, the spherical Dirichlet-Voronoi cell is defined by

$$D'(\mathbf{x}_i) = \{\mathbf{x} \in \mathrm{bd}(S_n) : \|\mathbf{x}, \mathbf{x}_i\|_g \le \|\mathbf{x}, \mathbf{x}_j\|_g, \ j = 1, 2, \ldots, m\}.$$

In other words,

$$D'(\mathbf{x}_i) = \bigcap_{j=1}^{m} H_{ij},$$

where

$$H_{ij} = \{\mathbf{x} \in \mathrm{bd}(S_n) : \langle \mathbf{x}, \mathbf{x}_i - \mathbf{x}_j \rangle \ge 0\}.$$

From the definition of $D'(\mathbf{x}_i)$, it is easy to see that

$$\Omega(\mathbf{x}_i, \phi) \subset D'(\mathbf{x}_i)$$

for $i = 1, 2, \ldots, m$. Thus, if we can prove

$$\frac{\mu(n, \phi)}{s(D'(\mathbf{x}_i))} \le \sigma_n(\phi)$$

for all indices $i = 1, 2, \ldots, m$, we can deduce (7.18) immediately.

In $\mathrm{bd}(S_n)$, $n \ge 3$, let $g_i(\phi)$ be the geodesic distance from the center of an i-dimensional regular spherical simplex of side length 2ϕ to its vertices, and let $g(\phi)$ be the geodesic distance from the center of a regular spherical cross polytope of side length 2ϕ to its vertices. It is easy to see that

$$\phi = g_1(\phi) < g_2(\phi) < \cdots < g_{n-1}(\phi) < g(\phi).$$

Let $X = \{\mathbf{x}_1, \mathbf{x}_2, \ldots, \mathbf{x}_{m(n,\phi)}\}$ be a discrete set of points in $\mathrm{bd}(S_n)$ such that the caps $\Omega(\mathbf{x}_i, \phi)$, $i = 1, 2, \ldots, m(n, \phi)$, form a cap packing in $\mathrm{bd}(S_n)$. We introduce two simple results concerning the geometry of the spherical Dirichlet-Voronoi cells $D'(\mathbf{x}_i)$.

Lemma 7.3. *Let F_{n-i-1} be an $(n - i - 1)$-dimensional common face of the $i + 1$ spherical Dirichlet-Voronoi cells $D'(\mathbf{x}_1), D'(\mathbf{x}_2), \ldots, D'(\mathbf{x}_{i+1})$. Let S'_{n-i-1} be the spherical subspace of $\mathrm{bd}(S_n)$ determined by F_{n-i-1}. Then, for $j = 1, 2, \ldots, i + 1$,*

$$\min_{\mathbf{x} \in S'_{n-i-1}} \|\mathbf{x}, \mathbf{x}_j\|_g \ge g_i(\phi),$$

where equality holds if and only if the $i + 1$ caps $\Omega(x_j, \phi)$ mutually touch one another.

Proof. Let x be an arbitrary point of S'_{n-i-1}. Then the relations

$$\|x, x_j\|_g = \varphi, \quad j = 1, 2, \ldots, i+1,$$

hold simultaneously for a certain positive number φ. Let S^*_{i+1} be the spherical subspace determined by the $i + 1$ points $x_1, x_2, \ldots, x_{i+1}$. Observing the distribution of the $i + 1$ points in the spherical cap $S^*_{i+1} \cap \Omega(x, \varphi)$, it follows from routine argument that

$$\varphi \geq g_i(\phi),$$

where equality holds if and only if the $i + 1$ caps $\Omega(x_j, \phi)$ mutually touch one another. Thus, Lemma 7.3 is proved. $\square$

Lemma 7.4. *Let F_{n-i-1} be an $(n - i - 1)$-dimensional common face of $i + 1$ spherical Dirichlet-Voronoi cells $D'(x_1)$, $D'(x_2)$, $\ldots$, $D'(x_{i+1})$; let S'_{n-i-1} be the spherical subspace determined by F_{n-i-1}; and let $S^\star$ be the spherical subspace*

$$\{x \in \text{bd}(S_n): \ \|x, x_1\|_g = \|x, x_{i+2}\|_g\}.$$

Suppose that y is a point of S'_{n-i-1} such that

$$\|x_1, y\|_g = \min_{x \in S'_{n-i-1}} \|x_1, x\|_g$$

and suppose that $\widehat{x_1 y} \cap S^\star \neq \emptyset$. Then

$$\|x_1, y\|_g \geq g(\phi).$$

Proof. Let S^*_j be the spherical subspace determined by $x_1, x_2, \ldots, x_j$. Obviously, $y \in S^*_{i+1}$. For convenience, we write

$$\Omega = \Omega(y, \|x_1, y\|_g) \cap S^*_{i+2}.$$

It is easy to see that the $i + 1$ points $x_1, x_2, \ldots, x_{i+1}$ lie on the relative boundary of the cap Ω. Since $\widehat{x_1 y} \cap S^\star \neq \emptyset$, it follows from Figure 7.1 that $x_{i+2} \in \Omega$. Let x'_{i+2} be a point on the relative boundary of Ω such that $\widehat{x_{i+2} x'_{i+2}}$ is orthogonal to S^*_{i+1}. Then we have

$$\|x_j, x'_{i+2}\|_g \geq \|x_j, x_{i+2}\| \geq 2\phi$$

for $j = 1, 2, \ldots, i + 1$.

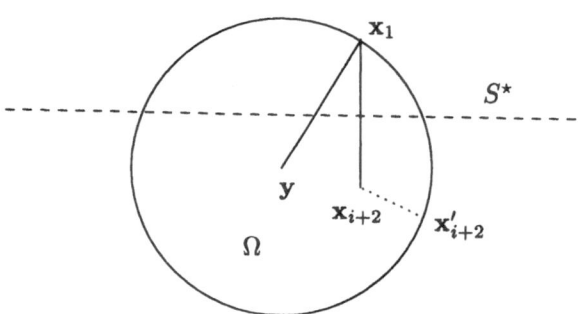

Figure 7.1

Now, the $i + 2$ points $\mathbf{x}_1$, $\mathbf{x}_2$, ..., $\mathbf{x}_{i+1}$, and $\mathbf{x}'_{i+2}$ lie on the boundary of a hemicap of Ω bounded by S^*_{i+1}. From elementary geometry we know that in this situation there are two points, $\mathbf{x}_1$ and $\mathbf{x}_2$ say, such that the spherical angle between $\widehat{\mathbf{x}_1\mathbf{y}}$ and $\widehat{\mathbf{y}\mathbf{x}_2}$ is less than or equal to $\pi/2$. Thus, we have

$$\|\mathbf{x}_1, \mathbf{y}\|_g \geq g(\phi).$$

Lemma 7.4 is proved.   □

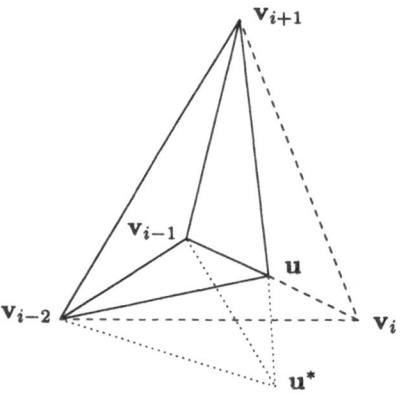

Figure 7.2

In $\mathrm{bd}(S_n)$, as usual, we say that the spherical segments $\widehat{\mathbf{x}\mathbf{y}}$ and $\widehat{\mathbf{y}\mathbf{z}}$ are orthogonal at $\mathbf{y}$ if their projections onto the tangent hyperplane of $\mathrm{bd}(S_n)$ at $\mathbf{y}$ are orthogonal. Then, a spherical simplex $T = \mathbf{v}_1\mathbf{v}_2\ldots\mathbf{v}_n$, where $\|\mathbf{v}_1, \mathbf{v}_i\|_g < \pi/2$ for $i = 1, 2, \ldots, n$, is called an *orthosimplex* if $\widehat{\mathbf{v}_i\mathbf{v}_j}$ and $\widehat{\mathbf{v}_j\mathbf{v}_k}$ are pairwise orthogonal whenever $1 \leq i < j < k \leq n$. More generally, if T is an orthosimplex and $\mathbf{u} \in \widehat{\mathbf{v}_{i-1}\mathbf{v}_i}$, we call the spherical simplex

$$T_{n,i,\mathbf{u}} = \mathbf{v}_1 \ldots \mathbf{v}_{i-1}\mathbf{u}\mathbf{v}_{i+1} \ldots \mathbf{v}_n$$

a *coorthosimplex*. It is easy to see that the coorthosimplex can be obtained from another orthosimplex $T^* = \mathbf{v}_n \ldots \mathbf{v}_{i+1}\mathbf{u}^*\mathbf{v}_{i-1} \ldots \mathbf{v}_1$ such that $\mathbf{u} \in \widehat{\mathbf{u}^*\mathbf{v}_{i+1}}$ (see Figure 7.2). For convenience, we call $\mathbf{u}^*$ and $\mathbf{u}^{**} = \mathbf{v}_i$ the lower *orthopoint* and upper orthopoint of the vertex $\mathbf{u}$ of $T_{n,i,\mathbf{u}}$, respectively. Clearly, $\mathbf{v}_i^* = \mathbf{v}_i^{**} = \mathbf{v}_i$ if T is an orthosimplex.

In order to prove a basic result about orthosimplices, we introduce a couple of basic properties of coorthosimplices.

Assertion 7.1. *Let $\mathbf{p}$ be a point of $T_{n,i,\mathbf{u}}$ such that $\mathbf{p} \in \widehat{\mathbf{v}_{i-1}\mathbf{u}}$, and let $\mathbf{p}^*$ and $\mathbf{p}^{**}$ be the lower orthopoint and upper orthopoint of the vertex $\mathbf{p}$ of the subcoorthosimplex $\mathbf{v}_1 \ldots \mathbf{v}_{i-2}\mathbf{p}\mathbf{v}_{i+1} \ldots \mathbf{v}_n$, respectively. When we move $\mathbf{p}$ along $\widehat{\mathbf{v}_{i-1}\mathbf{u}}$ from $\mathbf{v}_{i-1}$ to $\mathbf{u}$, both $\|\mathbf{v}_1,\mathbf{p}^*\|_g$ and $\|\mathbf{v}_1,\mathbf{p}^{**}\|_g$ increase.*

Proof. We examine only the $\|\mathbf{v}_1,\mathbf{p}^*\|_g$ case.

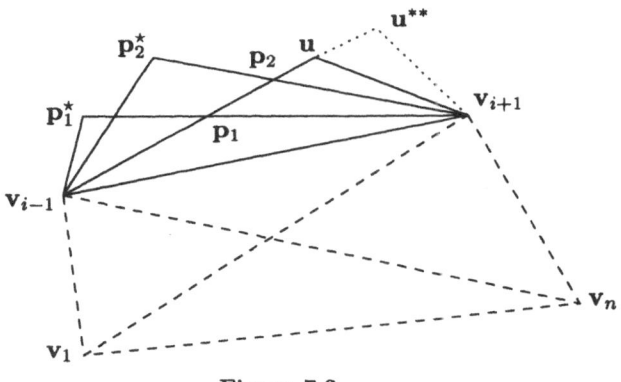

Figure 7.3

When $i \le 2$ or $i = n$, the assertion is trivial. Otherwise, let $\mathbf{p}_1$ and $\mathbf{p}_2$ be two distinct points of the spherical segment $\widehat{\mathbf{v}_{i-1}\mathbf{u}}$ such that

$$\|\mathbf{v}_{i-1},\mathbf{p}_1\|_g < \|\mathbf{v}_{i-1},\mathbf{p}_2\|_g. \tag{7.20}$$

By elementary arguments, it can be shown that

$$\mathbf{p}_j^* = \mathbf{p}_j^*, \quad j = 1,\, 2. \tag{7.21}$$

Then, it follows from the definition of the lower orthopoint that $\mathbf{v}_{i-1}\mathbf{p}_1^*\mathbf{v}_{i+1}$ and $\mathbf{v}_{i-1}\mathbf{p}_2^*\mathbf{v}_{i+1}$ are right spherical triangles lying in the spherical subspace S determined by $\mathbf{v}_{i-1}$, $\mathbf{u}^{**}$, and $\mathbf{v}_{i+1}$, with common hypotenuse $\widehat{\mathbf{v}_{i-1}\mathbf{v}_{i+1}}$ (see Figure 7.3). Therefore, it follows from (7.20) and (7.21) that

$$\|\mathbf{v}_{i-1},\mathbf{p}_1^*\|_g < \|\mathbf{v}_{i-1},\mathbf{p}_2^*\|_g. \tag{7.22}$$

Observe that $\widehat{\mathbf{v}_1\mathbf{v}_{i-1}}$ is orthogonal to S. Then (7.22) implies

$$\|\mathbf{v}_1,\mathbf{p}_1^*\|_g < \|\mathbf{v}_1,\mathbf{p}_2^*\|_g.$$

Thus, the considered case is proved.   □

Assertion 7.2. *Let T_{n,i,u_1} and T_{n,i,u_2} be coorthosimplices in* $\mathrm{bd}(S_n)$ *lying in the same side of the spherical subspace determined by* $v_1, \ldots, v_{i-1}, v_{i+1},$ $\ldots, v_n$. *Suppose that*

$$\|v_1, u_1^*\|_g < \|v_1, u_2^*\|_g \quad and \quad \|v_1, u_1^{**}\|_g < \|v_1, u_2^{**}\|_g.$$

Then, the spherical segments $\widehat{u_1 v_{i+1}}$ *and* $\widehat{v_{i-1} u_2}$ *intersect one another at a point in the relative interior of both.*

Proof. Since both the spherical subspaces S and S', determined by v_{i-1}, u_1^*, u_1^{**}, and v_{i+1}, and by v_{i-1}, u_2^*, u_2^{**}, and v_{i+1}, respectively, are orthogonal to both spherical simplices $v_1 v_2 \ldots v_{i-1}$ and $v_{i+1} v_{i+2} \ldots v_n$, it follows that

$$S = S'.$$

Then, the assertion follows from Assertion 7.1 and routine argument (see Figure 7.4).   □

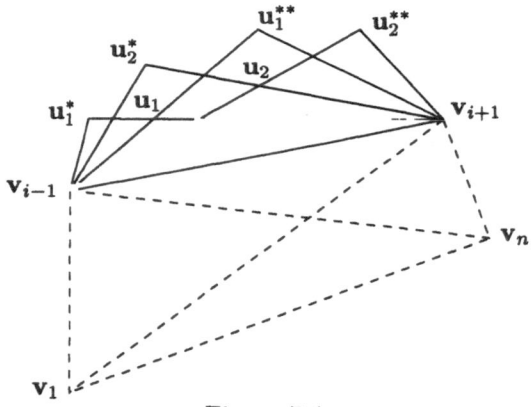

Figure 7.4

Let $T_n = v_1 v_2 \ldots v_n$ be a spherical simplex in $\mathrm{bd}(S_n)$ with $s(T_n) \neq 0$. Then, we define

$$\delta(T_n, \phi) = \frac{s(\Omega(v_1, \phi) \cap T_n)}{s(T_n)}. \tag{7.23}$$

Also, if T_m is a spherical simplex in $\mathrm{bd}(S_n)$ with $m < n$, for $i \leq m$ we define

$$\delta(T_m, v_i, \phi) = \lim_{\substack{v_j \to v_i \\ m < j \leq n}} \delta(T_n, \phi). \tag{7.24}$$

In this case, letting V be the spherical set obtained by rotating T_m about the spherical subspace determined by $v_1, \ldots, v_{i-1}, v_{i+1}, \ldots, v_m$, by (7.23) and (7.24) it follows that

$$\delta(T_m, v_i, \phi) = \frac{s(\Omega(v_1, \phi) \cap V)}{s(V)}.$$

Now we state and prove the following basic lemma.

Lemma 7.5 (Böröczky [1]). *Let T_{m,i,u_1} and T_{m,i,u_2} be coorthosimplices in $\mathrm{bd}(S_n)$ such that*

$$\|v_1, u_1^*\|_g < \|v_1, u_2^*\|_g \quad and \quad \|v_1, u_1^{**}\|_g < \|v_1, u_2^{**}\|_g,$$

and let $\phi < \pi/4$ be a positive number such that $\Omega(v_1, \phi)$ intersects the spherical subspace S_j° in at most one point, where S_j° is determined by v_2, $\ldots$, v_{i-1}, u_j, v_{i+1}, $\ldots$, v_m. Then,

$$\delta(T_{m,i,u_1}, u_1, \phi) \geq \delta(T_{m,i,u_2}, u_2, \phi).$$

In particular, let $T = v_1 v_2 \ldots v_m$ and $T' = v_1 v_2' \ldots v_n'$ be orthosimplices such that

$$\|v_1, v_m\|_g \geq \|v_1, v_n'\|_g \quad and \quad \|v_1, v_j\|_g \geq \|v_1, v_j'\|_g$$

for $j = 2, 3, \ldots, m-1$, and let $\phi < \pi/4$ be a positive number such that $\phi \leq \|v_1, v_2'\|_g$. Then,

$$\delta(T, v_m, \phi) \leq \delta(T', \phi).$$

Proof. To prove the first part, we apply induction on m. It is obvious for $m = 2$. Supposing that it is true for $m < k \leq n$, we proceed to prove the case $m = k$.

Let p and q be points of the spherical segments $\widehat{u_1 v_{i+1}}$ and $\widehat{v_{i-1} u_2}$, respectively. For $k = n$, let $v = v(p)$ be the area of $T_{n,i,p}$, and let $w = w(q)$ be the area of $T_{n,i,q}$. If $k < n$, let $v(p)$ and $w(q)$ be the areas of the spherical bodies obtained by rotating $T_{k,i,p}$ and $T_{k,i,q}$, respectively, about the spherical subspace S determined by $v_1, \ldots, v_{i-1}, v_{i+1}, \ldots, v_k$. Then, we write

$$f_1(v) = \delta(v_1 \ldots v_{i-1} p v_{i+2} \ldots v_k, p, \phi)$$

and

$$f_2(w) = \delta(v_1 \ldots v_{i-2} q v_{i+1} \ldots v_k, q, \phi).$$

By Assertion 7.1 and the inductive assumption it follows that $f_1(v)$ is an increasing function of v, and $f_2(w)$ is a decreasing function of w.

Without loss of generality, we may assume that T_{k,i,u_1} and T_{k,i,u_2} lie in a common subspace on the same side of the subspace determined by $v_1, \ldots, v_{i-1}, v_{i+1}, \ldots, v_k$. First, we deal with the case $2 < i < k$. By Assertion 7.2, the two spherical segments $\widehat{v_{i-1} u_2}$ and $\widehat{u_1 v_{i+1}}$ intersect one another at a point w, say. It is easy to see that

$$\delta(T_{k,i,u_1}, u_1, \phi) = \frac{1}{v(u_1)} \int_0^{v(u_1)} f_1(v) dv,$$

$$\delta(T_{k,i,u_2}, u_2, \phi) = \frac{1}{w(u_2)} \int_0^{w(u_2)} f_2(w) dw,$$

and

$$\delta(T_{k,i,\mathbf{w}}, \mathbf{w}, \phi) = \frac{1}{v(\mathbf{w})} \int_0^{v(\mathbf{w})} f_1(v) dv$$

$$= \frac{1}{w(\mathbf{w})} \int_0^{w(\mathbf{w})} f_2(w) dw.$$

Since $f_1(v)$ is increasing and $f_2(w)$ is decreasing, we have

$$\delta(T_{k,i,\mathbf{u}_1}, \mathbf{u}_1, \phi) \geq \delta(T_{k,i,\mathbf{w}}, \mathbf{w}, \phi) \geq \delta(T_{k,i,\mathbf{u}_2}, \mathbf{u}_2, \phi).$$

The other cases $i = 2$ and $i = k$ are routine. Thus, we have proved the first part of our lemma.

Now we deal with the second part. If $m = n$, letting

$$T_i'' = \mathbf{v}_1 \mathbf{v}_2'' \ldots \mathbf{v}_i'' \mathbf{v}_{i+1} \ldots \mathbf{v}_n$$

be an orthosimplex such that

$$\|\mathbf{v}_1, \mathbf{v}_j''\|_g = \|\mathbf{v}_1, \mathbf{v}_j'\|_g, \quad j = 2, 3, \ldots, i,$$

by the first assertion of this lemma it follows that

$$\delta(T, \phi) \leq \delta(T_2'', \phi) \leq \ldots \leq \delta(T_n'', \phi) = \delta(T', \phi). \tag{7.25}$$

When $m < n$, by choosing $\mathbf{v}_{m+1}, \mathbf{v}_{m+2}, \ldots, \mathbf{v}_n$ near to $\mathbf{v}_m$ and applying (7.25), it follows that

$$\delta(T, \mathbf{v}_m, \phi) \leq \delta(T', \phi).$$

Lemma 7.5 is proved.   □

With these preparations, we can prove the following key lemma of this section.

Lemma 7.6 (Böröczky [1]). *Let $\phi < \pi/4$ be a positive number, and let $X = \{\mathbf{x}_1, \mathbf{x}_2, \ldots, \mathbf{x}_{m(n,\phi)}\}$ be a discrete set in $\mathrm{bd}(S_n)$ such that the corresponding caps $\Omega(\mathbf{x}_i, \phi)$ together form a packing in $\mathrm{bd}(S_n)$. Then,*

$$\frac{s(\Omega(\mathbf{x}_i, \phi))}{s(D'(\mathbf{x}_i))} \leq \sigma_n(\phi)$$

holds for every point $\mathbf{x}_i \in X$.

Proof. Without loss of generality, we consider $\Omega(\mathbf{x}_1, \phi)$ and $D'(\mathbf{x}_1)$ and, for convenience, abbreviate them to Ω and D, respectively. Let F_{n-i-1} be an $(n-i-1)$-dimensional face of D. By Lemma 7.3 it follows that

$$\|\mathbf{x}_1, F_{n-i-1}\|_g \geq g_i(\phi).$$

Let Ω' be the cap $\{\mathbf{x} \in \text{bd}(S_n) : \|\mathbf{x}_1, \mathbf{x}\|_g < g_{n-1}(\phi)\}$, let S_{n-i-1}° be the spherical subspace determined by F_{n-i-1}, and let $\mathbf{p}_i$ be the point of S_{n-i-1}° satisfying

$$\|\mathbf{x}_1, \mathbf{p}_i\|_g = \min\left\{ \|\mathbf{x}_1, \mathbf{x}\|_g : \mathbf{x} \in S_{n-i-1}^{\circ} \right\}.$$

First, we assert that $\mathbf{p}_i \in F_{n-i-1}$ if $\Omega' \cap F_{n-i-1} \neq \emptyset$. If, on the contrary, $\mathbf{p}_i \notin F_{n-i-1}$, we proceed to show $\Omega' \cap F_{n-i-1} = \emptyset$. Assume that F_{n-i-1} is a common face of $i+1$ spherical Dirichlet-Voronoi cells, $D'(\mathbf{x}_1)$, $D'(\mathbf{x}_2)$, ..., $D'(\mathbf{x}_{i+1})$. There is an $(n-i-2)$-dimensional face F_{n-i-2} of F_{n-i-1} such that the corresponding subspace S_{n-i-2}° separates $\mathbf{p}_i$ from F_{n-i-1} (in S_{n-i-1}°). Assume that F_{n-i-2} is a common face of $D'(\mathbf{x}_1)$, $D'(\mathbf{x}_2)$, ..., $D'(\mathbf{x}_{n+2})$ and write

$$S = \{\mathbf{x} \in \text{bd}(S_n) : \|\mathbf{x}_1, \mathbf{x}\|_g = \|\mathbf{x}_{i+2}, \mathbf{x}\|_g\}.$$

Then S separates $\mathbf{p}_i$ from D, which means that S intersects the spherical segment $\widehat{\mathbf{x}_1 \mathbf{p}_i}$. Thus, by Lemma 7.4 it follows that

$$\|\mathbf{x}_1, \mathbf{p}_i\|_g \geq g(\phi),$$

which implies that $\Omega' \cap F_{n-i-1} = \emptyset$. The first assertion follows from this contradiction.

Second, we proceed to show

$$\frac{s(\Omega)}{s(D \cap \Omega')} \leq \sigma_n(\phi). \tag{7.26}$$

To do this, we decompose $D \cap \Omega'$ into smaller cells and consider the local densities associated to them. For convenience, denote the set formed by adjoining the points of $\text{conv}\{\mathbf{v}_1, \mathbf{v}_2, ..., \mathbf{v}_j\}$ to a set Y by $[\mathbf{v}_1 \mathbf{v}_2 ... \mathbf{v}_j \bowtie Y]$, and denote the relative boundary of Ω' by $\text{bd}(\Omega')$. The decomposition is defined by induction as follows. To start, write $\mathbf{v}_1 = \mathbf{x}_1$ and divide $D \cap \Omega'$ into a set $[\mathbf{v}_1 \bowtie D \cap \text{bd}(\Omega')]$ and sets $[\mathbf{v}_1 \bowtie F_{n-2} \cap \Omega']$, where F_{n-2} runs over the $(n-2)$-dimensional faces of D. Suppose that in the kth step $(1 \leq k < n-1)$ we have decomposed $D \cap \Omega'$ into sets of type $[\mathbf{v}_1 ... \mathbf{v}_j \bowtie F_{n-j} \cap \text{bd}(\Omega')]$ for $j = 1, 2, ..., k-1$ (where $F_{n-1} = D$), and sets of type $[\mathbf{v}_1 ... \mathbf{v}_{k-1} \bowtie F_{n-k} \cap \Omega']$. We do the following in the next step: Consider one set $[\mathbf{v}_1 ... \mathbf{v}_{k-1} \bowtie F_{n-k} \cap \Omega']$ of the second type and take $\mathbf{v}_k$ to be the point $\mathbf{p}_{k-1}$ corresponding to S_{n-k}°. By the first assertion it follows that $\mathbf{v}_k \in F_{n-k} \cap \Omega'$ if $F_{n-i} \cap \Omega' \neq \emptyset$. Then, we divide $[\mathbf{v}_1 ... \mathbf{v}_{k-1} \bowtie F_{n-k} \cap \Omega']$ into a subset $[\mathbf{v}_1 ... \mathbf{v}_k \bowtie F_{n-k} \cap \text{bd}(\Omega')]$ and subsets $[\mathbf{v}_1 ... \mathbf{v}_k \bowtie F_{n-k-1} \cap \Omega']$, where F_{n-k-1} runs over all $(n-k-1)$-dimensional faces of F_{n-k}. By Lemma 7.3, all vertices of D lie outside of Ω', which means that there is no set of type $[\mathbf{v}_1 ... \mathbf{v}_{n-1} \bowtie F_0 \cap \Omega']$. Thus, the process ends at the $(n-1)$th step. As a result, we have decomposed $D \cap \Omega'$ into sets of type $[\mathbf{v}_1 ... \mathbf{v}_j \bowtie F_{n-j} \cap \text{bd}(\Omega')]$, $j = 1, 2, ..., n-1$ (see Figure 7.5). In particular, the sets of type $[\mathbf{v}_1 ... \mathbf{v}_{n-1} \bowtie F_0 \cap \text{bd}(\Omega')]$ are orthosimplices.

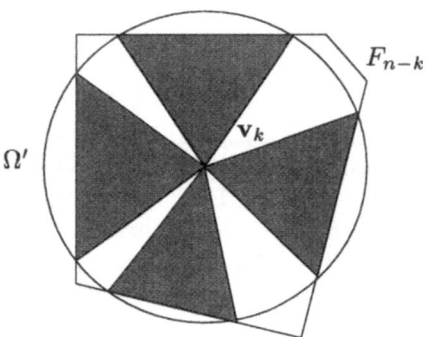

Figure 7.5

Third, writing

$$\Pi_j = [\mathbf{v}_1 \ldots \mathbf{v}_j \bowtie F_{n-j} \cap \mathrm{bd}(\Omega')],$$

we proceed to show that

$$\delta(\Pi_j, \phi) \leq \sigma_n(\phi). \tag{7.27}$$

Let $\Omega(\mathbf{u}_i, \phi)$, $i = 1, 2, \ldots, n$, be n mutually touching caps, and let $T^\star$ be the regular spherical simplex spanned by the centers of these caps. $T^\star$ can be divided into $n!$ congruent orthosimplices. One of them, say $T = \mathbf{t}_1 \mathbf{t}_2 \ldots \mathbf{t}_n$, is obtained by taking $\mathbf{t}_1$ to be a vertex of $T^\star$, and when $\mathbf{t}_1$, $\mathbf{t}_2$, $\ldots$, $\mathbf{t}_{i-1}$ have been chosen, taking $\mathbf{t}_i$ to be the centroid of one of the i-dimensional faces of $T^\star$ that contains $\mathbf{t}_1$, $\mathbf{t}_2$, $\ldots$, $\mathbf{t}_{i-1}$. Then, we have

$$\delta(T, \phi) = \sigma_n(\phi). \tag{7.28}$$

Consider the set Π_j and let $\mathbf{v}_{j+1}$ be an arbitrary point of $F_{n-j} \cap \mathrm{bd}(\Omega')$. It is obvious that

$$\delta(\Pi_j, \phi) \leq \delta(\mathbf{v}_1 \mathbf{v}_2 \ldots \mathbf{v}_{j+1}, \mathbf{v}_{j+1}, \phi). \tag{7.29}$$

By Lemma 7.3 it follows that $\|\mathbf{v}_1, \mathbf{v}_i\|_g \geq g_{i-1}(\phi)$ for $i = 1, 2, \ldots, j$. Also, it is obvious that $\|\mathbf{v}_1, \mathbf{v}_{j+1}\|_g = g_{n-1}(\phi)$. On the other hand, by definition, we have $\|\mathbf{t}_1, \mathbf{t}_i\|_g = g_{i-1}(\phi)$ for $i = 2, \ldots, n$. Thus, (7.29), Lemma 7.5, and (7.28) imply that

$$\delta(\Pi_j, \phi) \leq \delta(\mathbf{v}_1 \mathbf{v}_2 \ldots \mathbf{v}_{j+1}, \mathbf{v}_{j+1}, \phi)$$
$$\leq \delta(T, \phi) = \sigma_n(\phi),$$

which proves (7.27).

Obviously, (7.27) implies (7.26), and therefore the lemma is proved. $\square$

Proof of Conjecture 7.1. Let $\phi < \pi/4$ be a positive number and let $X = \{\mathbf{x}_1, \mathbf{x}_2, \ldots, \mathbf{x}_{m(n,\phi)}\}$ be a set of points in bd(S_n) such that the caps $\Omega(\mathbf{x}_i, \phi)$, $\mathbf{x}_i \in X$, form a packing in bd(S_n). By Lemma 7.6 it follows that

$$\frac{m(n,\phi)\mu(n,\phi)}{n\omega_n} = \frac{\sum_{i=1}^{m(n,\phi)} s(\Omega(\mathbf{x}_i,\phi))}{\sum_{i=1}^{m(n,\phi)} s(D'(\mathbf{x}_i))}$$

$$\leq \max_{1 \leq i \leq m(n,\phi)} \frac{s(\Omega(\mathbf{x}_i,\phi))}{s(D'(\mathbf{x}_i))}$$

$$\leq \sigma_n(\phi).$$

Let $T = \mathbf{v}_1\mathbf{v}_2\ldots\mathbf{v}_n$ be a regular spherical simplex in bd(S_n) of geodesic side length 2ϕ and corresponding dihedral angle 2α. Thus,

$$m(n,\phi) \leq \frac{n\omega_n\sigma_n(\phi)}{\mu(n,\phi)} = \frac{n^2\omega_n s(T \cap \Omega(\mathbf{v}_1,\phi))}{s(T)s(\Omega(\mathbf{v}_1,\phi))}. \tag{7.30}$$

From the definition of Schläfli's function, we have

$$s(T) = \frac{n!n\omega_n}{2^n}F_n(\alpha). \tag{7.31}$$

Also, by projecting T and $\Omega(\mathbf{v}_1,\phi)$ onto the tangent hyperplane of S_n at $\mathbf{v}_1$, it is easy to see that

$$\frac{s(T \cap \Omega(\mathbf{v}_1,\phi))}{s(\Omega(\mathbf{v}_1,\phi))} = \frac{(n-1)!}{2^{n-1}}F_{n-1}(\alpha). \tag{7.32}$$

Then, (7.30), (7.31), and (7.32) together yield

$$m(n,\phi) \leq \frac{n\omega_n\sigma_n(\phi)}{\mu(n,\phi)} = \frac{2F_{n-1}(\alpha)}{F_n(\alpha)}. \tag{7.33}$$

In order to determine the relationship between ϕ and α, we consider a special simplex T. Let b and d be suitable numbers such that T is determined by vertices $\mathbf{v}_1 = (b+d, b, \ldots, b)$, $\mathbf{v}_2 = (b, b+d, b, \ldots, b)$, $\ldots$, $\mathbf{v}_n = (b, b, \ldots, b, b+d)$. Then, it follows from elementary geometry that

$$\cos 2\phi = \frac{\langle \mathbf{v}_1, \mathbf{v}_2 \rangle}{\|\mathbf{v}_1\|^2} = \frac{2bd + nb^2}{d^2 + 2bd + nb^2}$$

and therefore

$$\sec 2\phi = \frac{d^2}{2bd + nb^2} + 1. \tag{7.34}$$

On the other hand, the internal angle 2α between two of the n bounding hyperplanes

$$H_i = \left\{ \mathbf{x} : \langle \mathbf{u}_i, \mathbf{x} \rangle = (nb+d)x^i - b\sum_{j=1}^{n} x^j = 0 \right\},$$

where $\mathbf{u}_i$ are suitable vectors, is given by

$$\cos 2\alpha = \frac{\langle \mathbf{u}_1, \mathbf{u}_2 \rangle}{\|\mathbf{u}_1\|^2} = \frac{2bd + nb^2}{d^2 + (n-1)(2bd + nb^2)}.$$

Then, by (7.34) we have

$$\sec 2\alpha = \frac{d^2}{2bd + nb^2} + n - 1 = \sec 2\phi + n - 2. \qquad (7.35)$$

Conjecture 7.1 follows from (7.33) and (7.35). $\square$

Proof of Theorem 7.2. Using (7.19), (7.18), and Lemma 7.2, since $\sec 2\alpha = n$ when $\phi = \pi/6$, it follows that

$$k(S_n) = m(n, \pi/6) \leq \frac{2F_{n-1}(\alpha)}{F_n(\alpha)} \sim \frac{\sqrt{\pi n^3}}{e\sqrt{2}} 2^{0.5n}.$$

Theorem 7.2 is proved. $\square$

7.4. Claude Ambrose Rogers

Claude Ambrose Rogers was born in Cambridge in 1920, but his family soon moved to north London. He went to boarding school in Berkhamsted, Herts, and then in 1938 he studied mathematics at University College London, being evacuated to Bangor for the first year of the war. His Ph.D thesis on divergent series was written under the supervision of R.G. Cooke of Birkbeck College and L.S. Bosanquet of University College London and accepted in 1949. In 1946 he joined the mathematical faculty of University College London. During that time, influenced by H. Davenport, he became interested in packing and covering, to which he has made contributions of fundamental importance.

In 1954 Rogers became the Mason professor of pure mathematics at the University of Birmingham. Four years later, when Davenport left London for Cambridge, Professor Rogers succeeded him as the Astor professor of pure mathematics at University College London, where he worked until his retirement in 1986.

Besides Theorem 3.2, Theorem 3.3, Theorem 3.4, and Theorem 7.1, Professor Rogers has made important contributions to the geometry of numbers, convex geometry, and geometric measure theory. For example, in 1957 in a joint paper he and G.C. Shephard proved that for every n-dimensional convex body K,

$$2^n \leq \frac{v(D(K))}{v(K)} \leq \binom{2n}{n},$$

where the upper bound can be realized if and only if K is a simplex and the lower bound can be realized if and only if K is centrally symmetric. In 1975 in another joint paper he and D.G. Larman proved that when $n \geq 12$, there exist n-dimensional centrally symmetric convex bodies C_1 and C_2 centered at $\mathbf{o}$ such that

$$v(C_1) > v(C_2)$$

and

$$s(C_1 \cap H) < s(C_2 \cap H)$$

for every hyperplane H that contains $\mathbf{o}$. The first example completely solves a problem of W. Blaschke, and the second solves a problem of H. Busemann and C.M. Petty in the corresponding dimensions. He is also the author of the classic book *Packing and Covering*.

Professor Rogers has been a fellow of the Royal Society since 1959, and he was the president of the London Mathematical Society for the term 1970–1972. For his distinguished contributions, he was awarded a De Morgan medal by the London Mathematical Society in 1977.

8. Upper Bounds for the Packing Densities and the Kissing Numbers of Spheres III

8.1. Jacobi Polynomials

The *Jacobi polynomials* are a family of well-known special functions. They are defined, for $k = 0, 1, 2, \ldots$, as

$$P_k^{\alpha,\beta}(t) = \frac{1}{2^k} \sum_{i=0}^{k} \binom{k+\alpha}{i} \binom{k+\beta}{k-i} (t+1)^i (t-1)^{k-i},$$

where $\alpha > -1$ and $\beta > -1$ are parameters. These polynomials play an important role in obtaining the Kabatjanski-Levenštein upper bounds for $\delta(S_n)$ and $k(S_n)$. In this section we introduce some of their fundamental properties. Since these properties are well-known and easy to deduce (see Szegö [1]), we omit their proofs.

Assertion 8.1.
$$P_k^{\alpha,\beta}(1) = \binom{k+\alpha}{k}.$$

Assertion 8.2.
$$\int_{-1}^{1} P_k^{\alpha,\beta}(t) P_l^{\alpha,\beta}(t) (1-t)^\alpha (1+t)^\beta dt = \delta_{kl} \varpi(k, \alpha, \beta),$$

where δ_{kl} is the Kronecker symbol and

$$\varpi(k, \alpha, \beta) = \frac{2^{\alpha+\beta+1} \Gamma(k+\alpha+1) \Gamma(k+\beta+1)}{k!(2k+\alpha+\beta+1) \Gamma(k+\alpha+\beta+1)}.$$

Conversely, if $P_0(t)$, $P_1(t)$, $P_2(t)$, ... is a sequence of polynomials such that

$$\int_{-1}^{1} P_k(t)P_l(t)(1-t)^{\alpha}(1+t)^{\beta}dt = \delta_{kl}\varpi(k,\alpha,\beta),$$

where the degree of $P_k(t)$ is k, then

$$P_k(t) = \pm P_k^{\alpha,\beta}(t), \quad k = 0,\ 1,\ 2,\ \dots.$$

Assertion 8.3. *The Jacobi polynomials $y = P_k^{\alpha,\beta}(t)$ satisfy the differential equation*

$$(1-t^2)y'' + [\beta - \alpha - (\alpha+\beta+2)t]y' + k(k+\alpha+\beta+1)y = 0.$$

Assertion 8.4 (The Christoffel-Darboux Formula).

$$\frac{P_{k+1}^{\alpha,\beta}(t)P_k^{\alpha,\beta}(s) - P_k^{\alpha,\beta}(t)P_{k+1}^{\alpha,\beta}(s)}{t-s}$$

$$= \frac{(2k+\alpha+\beta+2)(2k+\alpha+\beta+1)}{2(k+1)(k+\alpha+\beta+1)}\sum_{i=0}^{k}\frac{\varpi(k,\alpha,\beta)}{\varpi(i,\alpha,\beta)}P_i^{\alpha,\beta}(t)P_i^{\alpha,\beta}(s).$$

Clearly, by taking $s \to t$ one obtains the following result.

Corollary 8.1.

$$\left(P_{k+1}^{\alpha,\beta}(t)\right)' P_k^{\alpha,\beta}(t) - P_{k+1}^{\alpha,\beta}(t)\left(P_k^{\alpha,\beta}(t)\right)' > 0.$$

Assertion 8.5. *Let $t_1(k,\alpha,\beta)$, $t_2(k,\alpha,\beta)$, ..., $t_k(k,\alpha,\beta)$ be the k zeros (in decreasing order) of the polynomial $P_k^{\alpha,\beta}(t)$. Then, for $k = 1, 2, \dots$ and $i = 1, 2, \dots, k$,*

$$t_{i+1}(k+1,\alpha,\beta) < t_i(k,\alpha,\beta) < t_i(k+1,\alpha,\beta) < 1.$$

Assertion 8.6. *When $\alpha = \beta = (n-3)/2$,*

$$P_k^{\alpha,\beta}(t)P_l^{\alpha,\beta}(t) = \sum_{i=0}^{k+l} a_i(k,l)P_i^{\alpha,\beta}(t)$$

holds for some suitable nonnegative coefficients $a_i(k,l)$.

8.2. Delsarte's Lemma

A *spherical code* ℵ of dimension n, cardinality m, and minimum angle φ is a set of m points of $\mathrm{bd}(S_n)$ such that

$$\langle \mathbf{x}, \mathbf{y} \rangle \leq \cos \varphi$$

holds for any two distinct points $\mathbf{x}$ and $\mathbf{y}$ of ℵ. In other words, the m caps $\Omega(\mathbf{x}, \varphi/2)$, $\mathbf{x} \in$ ℵ, form a packing in $\mathrm{bd}(S_n)$. For convenience, we call ℵ an $\{n, m, \varphi\}$ code. Then, we define $m[n, \varphi]$ to be the maximum size, m, of any such spherical code. Clearly,

$$m(n, \varphi/2) = m[n, \varphi]. \tag{8.1}$$

As an upper bound for $m[n, \varphi]$, we have the following remarkable result.

Lemma 8.1 (Delsarte [1]). *Write* $\alpha = (n-3)/2$. *Let*

$$f(t) = \sum_{i=0}^{k} f_i P_i^{\alpha,\alpha}(t)$$

be a real polynomial such that $f_0 > 0$, $f_i \geq 0$ *for* $i = 1, 2, \ldots, k$, *and* $f(t) \leq 0$ *for* $-1 \leq t \leq \cos \varphi$. *Then*

$$m[n, \varphi] \leq \frac{f(1)}{f_0}.$$

Let $\mathcal{L}$ be the linear space of square integrable real functions over $\mathrm{bd}(S_n)$. For two functions $f_1(\mathbf{x})$ and $f_2(\mathbf{x})$ of $\mathcal{L}$ we define the inner product

$$\langle f_1 \circ f_2 \rangle = \frac{1}{n\omega_n} \int_{\mathrm{bd}(S_n)} f_1(\mathbf{x}) f_2(\mathbf{x}) d\omega(\mathbf{x}),$$

where $\omega(\mathbf{x})$ indicates the surface element of $\mathrm{bd}(S_n)$ at $\mathbf{x}$. Let $\mathcal{P}_k$ be the subspace of $\mathcal{L}$ consisting of *spherical polynomials* of degree at most k, and let $\mathcal{H}_k$ be the subspace of $\mathcal{L}$ consisting of *homogeneous harmonic polynomials* of degree k. Here, we say $f(\mathbf{x})$ is a harmonic if

$$\Delta f = \frac{\partial^2 f}{\partial x_1^2} + \frac{\partial^2 f}{\partial x_2^2} + \cdots + \frac{\partial^2 f}{\partial x_n^2} = 0,$$

where Δ is the *Laplace operator*.

Assertion 8.7.
$$\mathcal{P}_k = \mathcal{H}_k \oplus \mathcal{H}_{k-1} \oplus \cdots \oplus \mathcal{H}_0.$$

Proof. A polynomial $p_k(\mathbf{x})$ in $\mathcal{P}_k$ is a linear combination of *monomials*

$$m(\mathbf{x}) = x_1^{k_1} x_2^{k_2} \cdots x_n^{k_n}$$

with

$$k_1 + k_2 + \cdots + k_n = l \le k.$$

Since $\mathbf{x} \in \mathrm{bd}(S_n)$, we have $\langle \mathbf{x}, \mathbf{x} \rangle = 1$ and therefore

$$m(\mathbf{x}) = \langle \mathbf{x}, \mathbf{x} \rangle^i m(\mathbf{x}).$$

Take $i = \lfloor (k - l)/2 \rfloor$. Then $\langle \mathbf{x}, \mathbf{x} \rangle^i m(\mathbf{x})$ is a homogeneous polynomial, of degree k if $k - l$ is even and of degree $k - 1$ if $k - l$ is odd. Denote the subspace of $\mathcal{L}$ consisting of homogeneous polynomials of degree k by $\mathcal{H}_k^*$. Then

$$p_k(\mathbf{x}) = h_k^*(\mathbf{x}) + h_{k-1}^*(\mathbf{x}),$$

where $h_k^*(\mathbf{x}) \in \mathcal{H}_k^*$ and $h_{k-1}^*(\mathbf{x}) \in \mathcal{H}_{k-1}^*$. Also, since

$$\langle h_k^* \circ h_{k-1}^* \rangle = (-1)^{2k-1} \langle h_k^* \circ h_{k-1}^* \rangle,$$

we have

$$\langle h_k^* \circ h_{k-1}^* \rangle = 0.$$

Thus,

$$\mathcal{P}_k = \mathcal{H}_k^* \oplus \mathcal{H}_{k-1}^*. \tag{8.2}$$

By routine argument it is easy to see that Δ is a linear operator on $\mathcal{H}_k^*$ with kernel $\mathcal{H}_k$ and image $\mathcal{H}_{k-2}^*$. On the other hand, it is well-known from real analysis (see Groemer [5]) that

$$\int_{\mathrm{bd}(S_n)} f(\mathbf{x}) \Delta g(\mathbf{x}) d\omega(\mathbf{x}) = \int_{\mathrm{bd}(S_n)} g(\mathbf{x}) \Delta f(\mathbf{x}) d\omega(\mathbf{x})$$

holds for every pair of twice continuously differentiable functions $f(\mathbf{x})$ and $g(\mathbf{x})$ on $\mathrm{bd}(S_n)$. In particular, for $h_k(\mathbf{x}) \in \mathcal{H}_k$ and $h_{k-2}^*(\mathbf{x}) \in \mathcal{H}_{k-2}^*$, we have

$$\langle h_k \circ h_{k-2}^* \rangle = \frac{1}{n\omega_n} \int_{\mathrm{bd}(S_n)} g_k^*(\mathbf{x}) \Delta h_k(\mathbf{x}) d\omega(\mathbf{x}) = 0,$$

where $g_k^*(\mathbf{x})$ is a suitable function satisfying

$$h_{k-2}^*(\mathbf{x}) = \Delta g_k^*(\mathbf{x}).$$

Thus, we have

$$\mathcal{H}_k^* = \mathcal{H}_k \oplus \mathcal{H}_{k-2}^*. \tag{8.3}$$

Applying (8.2) and (8.3) inductively, it follows that

$$\mathcal{P}_k = \mathcal{H}_k \oplus \mathcal{H}_{k-1} \oplus \cdots \oplus \mathcal{H}_0.$$

Assertion 8.7 is proved.   □

Next we introduce a special class of *spherical harmonics*, which play an important role in our considerations. Since $\mathcal{H}_k$ is a finite-dimensional linear space equipped with a nontrivial inner product, it follows from a fundamental result of functional analysis that for any real-valued linear functional $F(h_k)$ defined on $\mathcal{H}_k$ there is a unique $g \in \mathcal{H}_k$ such that

$$F(h_k) = \langle h_k \circ g \rangle.$$

In particular, for each point $\mathbf{x} \in \mathrm{bd}(S_n)$, by taking $F(h_k) = h_k(\mathbf{x})$ it is easy to see that there is a unique function $Q_k(\mathbf{x}, \mathbf{y}) \in \mathcal{H}_k$ such that

$$h_k(\mathbf{x}) = \frac{1}{n\omega_n} \int_{\mathrm{bd}(S_n)} h_k(\mathbf{y}) Q_k(\mathbf{x}, \mathbf{y}) d\omega(\mathbf{y}). \tag{8.4}$$

Usually, $Q_k(\mathbf{x}, \mathbf{y})$ is called a *zonal spherical harmonic.*

Assertion 8.8. *For every isometry σ of E^n that leaves the origin fixed,*

$$Q_k(\sigma\mathbf{x}, \sigma\mathbf{y}) = Q_k(\mathbf{x}, \mathbf{y}).$$

Proof. Assume that $\mathbf{x} = \sigma\mathbf{y} = \mathbf{y}A$, where A is an $n \times n$ matrix satisfying $AA' = I$, and assume $h_k(\mathbf{x}) \in \mathcal{H}_k$. Then, routine computation yields

$$\Delta h_k(\sigma\mathbf{y}) = \sum_{i,j} \frac{\partial^2 h_k}{\partial x_i \partial x_j} \sum_{k=1}^{n} a_{ik} a_{jk}$$

$$= \Delta h_k(\mathbf{x}) = 0. \tag{8.5}$$

This means that $h_k(\sigma\mathbf{x}) \in \mathcal{H}_k$ if $h_k(\mathbf{x}) \in \mathcal{H}_k$. In particular, $Q_k(\sigma\mathbf{x}, \sigma\mathbf{y})$, as a function of $\mathbf{y}$, is a spherical harmonic.

Write

$$J_\sigma = \frac{1}{n\omega_n} \int_{\mathrm{bd}(S_n)} h_k(\mathbf{y}) Q_k(\sigma\mathbf{x}, \sigma\mathbf{y}) d\omega(\mathbf{y}).$$

By substituting $\mathbf{z} = \sigma\mathbf{y}$ and using (8.4), we have

$$J_\sigma = \frac{1}{n\omega_n} \int_{\mathrm{bd}(S_n)} h_k(\sigma^{-1}\mathbf{z}) Q_k(\sigma\mathbf{x}, \mathbf{z}) d\omega(\mathbf{z})$$

$$= h_k(\sigma^{-1}\sigma\mathbf{x}) = h_k(\mathbf{x})$$

$$= \frac{1}{n\omega_n} \int_{\mathrm{bd}(S_n)} h_k(\mathbf{y}) Q_k(\mathbf{x}, \mathbf{y}) d\omega(\mathbf{y}).$$

Then, Assertion 8.8 follows from the uniqueness of $Q_k(\mathbf{x}, \mathbf{y})$. □

Assertion 8.9 (Addition Formula). *Let $h_{k,1}(\mathbf{x})$, $h_{k,2}(\mathbf{x})$, $\ldots$, $h_{k,d(k)}(\mathbf{x})$ be an orthonormal basis of $\mathcal{H}_k$, where $d(k)$ is the dimension of $\mathcal{H}_k$. Then,*

$$Q_k(\mathbf{x}, \mathbf{y}) = \sum_{i=1}^{d(k)} h_{k,i}(\mathbf{x}) h_{k,i}(\mathbf{y}).$$

Proof. Any function $h_k(\mathbf{x}) \in \mathcal{H}_k$ can be expanded as

$$h_k(\mathbf{x}) = \sum_{i=1}^{d(k)} \langle h_k \circ h_{k,i} \rangle h_{k,i}(\mathbf{x}).$$

In particular, using (8.4),

$$Q_k(\mathbf{x}, \mathbf{y}) = \sum_{i=1}^{d(k)} \langle Q_k(\mathbf{x}, \mathbf{y}) \circ h_{k,i}(\mathbf{y}) \rangle h_{k,i}(\mathbf{y})$$

$$= \sum_{i=1}^{d(k)} h_{k,i}(\mathbf{x}) h_{k,i}(\mathbf{y}).$$

Assertion 8.9 is proved.   □

By Assertion 8.8 it follows that the value of $Q_k(\mathbf{x}, \mathbf{y})$ is determined by k and $\langle \mathbf{x}, \mathbf{y} \rangle$. So, for convenience, we write

$$G_k(\langle \mathbf{x}, \mathbf{y} \rangle) = Q_k(\mathbf{x}, \mathbf{y}).$$

Assertion 8.10. *Write $\alpha = (n-3)/2$. Then*

$$G_k(t) = c_k P_k^{\alpha, \alpha}(t)$$

holds for some positive number c_k.

Proof. By Assertion 8.7, for $k \neq l$, we have

$$J = \int_{\mathrm{bd}(S_n)} Q_k(\mathbf{x}, \mathbf{y}) Q_l(\mathbf{x}, \mathbf{y}) d\omega(\mathbf{y}) = 0.$$

Then, substituting $t = \langle \mathbf{x}, \mathbf{y} \rangle$ and

$$d\omega(\mathbf{y}) = (n-1)\omega_{n-1}(1-t^2)^{(n-3)/2} dt,$$

we obtain

$$J = (n-1)\omega_{n-1} \int_{-1}^{1} G_k(t) G_l(t) (1-t^2)^{(n-3)/2} dt = 0.$$

Thus, by the second part of Assertion 8.2 we have

$$G_k = c_k P_k^{\alpha, \alpha}(t), \quad k = 0, \ 1, \ \ldots . \tag{8.6}$$

To determine the sign of c_k, we observe that

$$G_k(1) = \frac{1}{n\omega_n} \int_{\mathrm{bd}(S_n)} G_k(1) d\omega(\mathbf{x})$$

$$= \frac{1}{n\omega_n} \int_{\mathrm{bd}(S_n)} \sum_{i=1}^{d(k)} h_{k,i}^2(\mathbf{x})d\omega(\mathbf{x})$$

$$= \sum_{i=1}^{d(k)} \frac{1}{n\omega_n} \int_{\mathrm{bd}(S_n)} h_{k,i}^2(\mathbf{x})d\omega(\mathbf{x})$$

$$= \sum_{i=1}^{d(k)} 1 = d(k). \tag{8.7}$$

Then, it follows from (8.6), (8.7), and Assertion 8.1 that $c_k > 0$ holds for $k = 0, 1, \ldots$. Assertion 8.10 is proved. $\qquad\square$

We can now proceed to prove Lemma 8.1.

Proof of Lemma 8.1. Let X be a discrete subset of $\mathrm{bd}(S_n)$ such that

$$\mathrm{card}\{X\} = m[n, \varphi],$$

and $\Omega(\mathbf{x}, \varphi/2)$, $\mathbf{x} \in X$, form a cap packing in $\mathrm{bd}(S_n)$. Then,

$$\langle \mathbf{x}, \mathbf{y} \rangle \leq \cos \varphi$$

holds for any two distinct points $\mathbf{x}$ and $\mathbf{y}$ of X.

Let

$$f(t) = \sum_{i=0}^{k} f_i P_i^{\alpha,\alpha}(t) \tag{8.8}$$

be a real polynomial such that $f_0 > 0$, $f_i \geq 0$ $(i = 1, 2, \ldots, k)$, and $f(t) \leq 0$ for $-1 \leq t \leq \cos \varphi$. For convenience, we apply

$$f(t) = \sum_{i=0}^{k} g_i G_i(t)$$

rather than (8.8), where $g_0 = f_0$, $G_0(t) = 1$, and $g_i = f_i/c_i \geq 0$ for $i = 1$, $2, \ldots, k$.

Write

$$J = \sum_{\mathbf{x}, \mathbf{y} \in X} f(\langle \mathbf{x}, \mathbf{y} \rangle).$$

Let W be the set of numbers w such that $w = \langle \mathbf{x}, \mathbf{y} \rangle$ holds for some pair of distinct points $\mathbf{x}$ and $\mathbf{y}$ of X. For $w \in W$ let $p(w)$ be the number of ordered pairs $(\mathbf{x}, \mathbf{y})$, $\mathbf{x}$, $\mathbf{y} \in X$, such that $\langle \mathbf{x}, \mathbf{y} \rangle = w$. Then, we have

$$J = m[n, \varphi]f(1) + \sum_{w \in W} p(w)f(w). \tag{8.9}$$

On the other hand, since $G_0(w) = Q_0(\mathbf{x}, \mathbf{y}) = 1$, the addition formula implies that

$$
\begin{aligned}
J &= \sum_{i=0}^{k} g_i \sum_{\mathbf{x},\,\mathbf{y} \in X} G_i(\langle \mathbf{x}, \mathbf{y} \rangle) = \sum_{i=0}^{k} g_i \sum_{\mathbf{x},\,\mathbf{y} \in X} Q_i(\mathbf{x}, \mathbf{y}) \\
&= g_0\, m[n,\varphi]^2 + \sum_{i=1}^{k} g_i \sum_{\mathbf{x},\,\mathbf{y} \in X} Q_i(\mathbf{x}, \mathbf{y}) \\
&= f_0\, m[n,\varphi]^2 + \sum_{i=1}^{k} g_i \sum_{j=1}^{d(i)} \left(\sum_{\mathbf{x} \in X} h_{i,j}(\mathbf{x}) \right)^2.
\end{aligned}
\tag{8.10}
$$

Comparing (8.9) and (8.10), since $f(w) \leq 0$ for $w \in W$, we have

$$
f(1)\, m[n, \varphi] \geq f_0\, m[n, \varphi]^2
$$

and therefore

$$
m[n, \varphi] \leq \frac{f(1)}{f_0}.
$$

This proves Lemma 8.1.   □

Remark 8.1. *This fundamental lemma was first proved by Delsarte in* [1] *and* [2]. *Alternative proofs can be found in Delsarte, Goethals, and Seidel* [1], *Kabatjanski and Levenštein* [1], *Odlyzko and Sloane* [1], *Lloyd* [1], *and Seidel* [1]. *Our proof follows Seidel's presentation.*

8.3. The Kabatjanski-Levenštein Upper Bounds for the Packing Densities and the Kissing Numbers of Spheres

For convenience, we write $\alpha = (n - 3)/2$ and denote by $\mathcal{F}(n, \varphi)$ the family of polynomials

$$
f(t) = \sum_{i=0}^{k} f_i P_i^{\alpha,\alpha}(t),
$$

such that $f_0 > 0$, $f_i \geq 0$ for $i = 1, 2, \ldots, k$, and $f(t) \leq 0$ for $-1 \leq t \leq \cos\varphi$. Then, we define

$$
m^*[n, \varphi] = \inf_{f \in \mathcal{F}(n,\varphi)} \frac{f(1)}{f_0}.
$$

It is obvious that

$$
m[n, \varphi] \leq m^*[n, \varphi].
\tag{8.11}
$$

As an upper bound for $m[n, \varphi]$, we have the following fundamental result.

Lemma 8.2 (Kabatjanski and Levenštein [1]). *When* $0 < \varphi < \pi/2$
and $n \to \infty$,

$$\frac{\log m[n, \varphi]}{n} \ll \frac{1 + \sin \varphi}{2 \sin \varphi} \log \frac{1 + \sin \varphi}{2 \sin \varphi} - \frac{1 - \sin \varphi}{2 \sin \varphi} \log \frac{1 - \sin \varphi}{2 \sin \varphi}.$$

The proof of this deep lemma requires some careful preparation.

Assertion 8.11. *Write* $s = \cos \varphi$ *and*

$$\tau = -\frac{k + 1}{k + \alpha + 1} \frac{P_{k+1}^{\alpha, \alpha}(s)}{P_k^{\alpha, \alpha}(s)}.$$

When $t_1(k, \alpha, \alpha) < s < t_1(k + 1, \alpha, \alpha)$ *(see Assertion 8.5),*

$$m^*[n, \varphi] \leq \frac{\binom{k + 2\alpha + 1}{k}(1 + \tau)^2}{(1 - s)\tau}.$$

Proof. Consider the polynomial

$$f(t) = \left(P_{k+1}^{\alpha, \alpha}(t) P_k^{\alpha, \alpha}(s) - P_k^{\alpha, \alpha}(t) P_{k+1}^{\alpha, \alpha}(s) \right) \frac{(k + \alpha + 1)(2k + 2\alpha + 1)}{(k + 1)(k + 2\alpha + 1)}$$

$$\times \sum_{i=0}^{k} \frac{\varpi(k, \alpha, \alpha)}{\varpi(i, \alpha, \alpha)} P_i^{\alpha, \alpha}(s) P_i^{\alpha, \alpha}(t) \tag{8.12}$$

of degree $2k + 1$. By Assertion 8.5 and our assumption it follows that

$$P_k^{\alpha, \alpha}(s) P_i^{\alpha, \alpha}(s) > 0 \quad \text{and} \quad P_{k+1}^{\alpha, \alpha}(s) P_i^{\alpha, \alpha}(s) < 0$$

for all $i \leq k$. Hence, by Assertions 8.1 and 8.6 we have $f_0 > 0$ and $f_i \geq 0$
for $i = 1, 2, \ldots, 2k + 1$. Also, by Assertion 8.4, we have

$$f(t) = \frac{\left(P_{k+1}^{\alpha, \alpha}(t) P_k^{\alpha, \alpha}(s) - P_k^{\alpha, \alpha}(t) P_{k+1}^{\alpha, \alpha}(s) \right)^2}{t - s} \tag{8.13}$$

and therefore $f(t) \leq 0$ when $-1 \leq t \leq s$. Thus,

$$f(t) \in \mathcal{F}(n, \varphi).$$

By Assertion 8.1 and (8.13) it follows that

$$f(1) = \frac{\left(P_k^{\alpha, \alpha}(s) \right)^2}{1 - s} \binom{k + \alpha}{k}^2 \left(\frac{k + \alpha + 1}{k + 1} - \frac{P_{k+1}^{\alpha, \alpha}(s)}{P_k^{\alpha, \alpha}(s)} \right)^2.$$

On the other hand, (8.12) and the first part of Assertion 8.2 together yield

$$
f_0\,\varpi(0,\alpha,\alpha) = \int_{-1}^{1} f(t)(1-t^2)^\alpha dt
$$

$$
= -\, P_{k+1}^{\alpha,\alpha}(s) P_k^{\alpha,\alpha}(s) \frac{(k+\alpha+1)(2k+2\alpha+1)\varpi(k,\alpha,\alpha)}{(k+1)(k+2\alpha+1)}.
$$

Hence

$$
f_0 = -P_{k+1}^{\alpha,\alpha}(s) P_k^{\alpha,\alpha}(s) \frac{(k+\alpha+1)(2k+2\alpha+1)\varpi(k,\alpha,\alpha)}{(k+1)(k+2\alpha+1)\varpi(0,\alpha,\alpha)}.
$$

Then, by Lemma 8.1, the definition of $\varpi(i,\alpha,\alpha)$, and routine computation we obtain

$$
m^*[n,\varphi] \le \frac{f(1)}{f_0} \le \frac{\binom{k+2\alpha+1}{k}(1+\tau)^2}{(1-s)\tau}.
$$

Assertion 8.11 is proved.   □

Assertion 8.12. *When* $s = \cos\varphi \le t_1(k,\alpha,\alpha)$,

$$
m^*[n,\varphi] \le \frac{4\binom{k+2\alpha+1}{k}}{1 - t_1(k+1,\alpha,\alpha)}.
$$

Proof. Regard τ as a function of s. When s decreases from $t_1(k+1,\alpha,\alpha)$ to $t_1(k,\alpha,\alpha)$, it follows from Corollary 8.1 that τ increases from 0 to ∞. So, $\tau = 1$ holds at a certain $s' = \cos\varphi'$ with

$$
t_1(k,\alpha,\alpha) < s' < t_1(k+1,\alpha,\alpha).
$$

Since $m^*[n,\varphi]$ is an increasing function of s, it follows from Assertion 8.11 that

$$
m^*[n,\varphi] \le m^*[n,\varphi'] \le \frac{4\binom{k+2\alpha+1}{k}}{1-s'}
$$

$$
\le \frac{4\binom{k+2\alpha+1}{k}}{1 - t_1(k+1,\alpha,\alpha)}.
$$

Assertion 8.12 is proved.   □

Assertion 8.13. *If* c *is a positive number such that*

$$
\lim_{k\to\infty} \frac{\alpha}{k} = \lim_{k\to\infty} \frac{n-3}{2k} = \frac{1}{2c}, \tag{8.14}
$$

then

$$
\lim_{k\to\infty} t_1(k,\alpha,\alpha) = \frac{2\sqrt{c(1+c)}}{1+2c}.
$$

Proof. Consider the function

$$y = (1 - t^2)^{(\alpha+1)/2} P_k^{\alpha,\alpha}(t), \tag{8.15}$$

which has the same zeros in $-1 < t < 1$ as $P_k^{\alpha,\alpha}(t)$. From Assertion 8.3, it can be deduced that this function satisfies the differential equation

$$y'' + g(t)y = 0, \tag{8.16}$$

where

$$g(t) = \frac{\eta(\gamma^2 - t^2)}{(1 - t^2)^2},$$

$$\eta = (k + \alpha)(k + \alpha + 1),$$

and

$$\gamma = \sqrt{[k(k + 2\alpha + 1) + \alpha + 1]/\eta}.$$

Since $\gamma < 1$, $g(t)$ is negative in $\gamma < t < 1$, and $y \to 0$ when $t \to 1$. Thus, by (8.16) it follows that y cannot vanish for $\gamma \le t < 1$. Consequently,

$$t_1(k, \alpha, \alpha) < \gamma. \tag{8.17}$$

On the other hand, for any positive number b the equation

$$y'' + b^2 y = 0$$

has a solution

$$y = \sin(bt + d),$$

where d is a constant, that has a zero between $t_1 = \gamma - 2\pi/b$ and $t_2 = \gamma - \pi/b$. If we choose b such that $t_1 \ge -1$ and $g(t) \ge b^2$ for $t_1 \le t \le t_2$, the function (8.15) will have a zero in this interval, and hence

$$t_1(k, \alpha, \alpha) \ge \gamma - \frac{2\pi}{b}. \tag{8.18}$$

In fact, taking

$$b = \left(\frac{\pi\gamma\eta}{4}\right)^{1/4},$$

it can be verified that $t_1 \ge -1$, $\gamma \ge 2\pi/b$, and

$$g(t) \ge \frac{(4\pi\gamma b - 4\pi^2)\eta}{[b^2 - (\gamma b)^2 + 2\pi\gamma b - \pi^2]^2} b^2 \ge b^2$$

holds for $t_1 \le t \le t_2$ when k is sufficiently large.

Then, by (8.17), (8.18), and routine computation we obtain

$$\lim_{k \to \infty} t_1(k, \alpha, \alpha) = \gamma = \frac{2\sqrt{c(1 + c)}}{1 + 2c}.$$

Assertion 8.13 is proved.   □

Proof of Lemma 8.2. Take

$$c = \frac{1 - \sin \varphi}{2 \sin \varphi} \tag{8.19}$$

and apply Assertion 8.13. This gives

$$\lim_{k \to \infty} t_1(k, \alpha, \alpha) = \cos \varphi.$$

Then, by (8.11) and Assertion 8.12 it follows that

$$m[n, \varphi] \le \frac{4\binom{k+2\alpha+1}{k}}{1 - t_1(k+1, \alpha, \alpha)}$$

$$\ll \frac{4}{1 - \cos \varphi} \binom{k + 2\alpha + 1}{k}. \tag{8.20}$$

Applying (8.14), Stirling's formula, and (8.19) to (8.20) we obtain

$$\frac{\log m[n, \varphi]}{n} \ll (1 + c)\log(1 + c) - c\log c$$

$$= \frac{1 + \sin \varphi}{2 \sin \varphi} \log \frac{1 + \sin \varphi}{2 \sin \varphi} - \frac{1 - \sin \varphi}{2 \sin \varphi} \log \frac{1 - \sin \varphi}{2 \sin \varphi}.$$

Lemma 8.2 is proved.   □

The special case $\varphi = \pi/3$ of Lemma 8.2 yields the following upper bound for the kissing numbers of spheres.

Theorem 8.1 (Kabatjanski and Levenštein [1]).

$$k(S_n) \le 2^{0.401n(1+o(1))}.$$

To get a corresponding upper bound for the packing densities of spheres, we need another lemma.

Lemma 8.3.

$$\delta(S_n) \le \left(\sin \frac{\varphi}{2} \right)^n m[n + 1, \varphi]$$

holds for a certain small φ.

Proof. Let r be a large number and let H be a hyperplane through the center of rS_{n+1}. In H we construct an n-dimensional packing of unit spheres with density $\delta(S_n)$. By shifting the packing slightly we may assume that the portion of H inside rS_{n+1} contains at least $r^n\delta(S_n)$ centers. We project these centers, perpendicularly to H, "upwards" onto $\mathrm{bd}(rS_{n+1})$. The Euclidean distance between the new points is at least 2, and their geodesic distance is at least φ, where

$$\sin \frac{\varphi}{2} = \frac{1}{r}.$$

Thus, we have

$$r^n \delta(S_n) \le m[n+1, \varphi]$$

and therefore

$$\delta(S_n) \le \left(\sin \frac{\varphi}{2} \right)^n m[n+1, \varphi].$$

Lemma 8.3 is proved. □

Remark 8.2. *This lemma first appeared in the appendix of the Russian edition of L. Fejes Tóth [9] (translated by I.M. Yaglom). Here, we follow the proof of Sloane [4].*

Theorem 8.2 (Kabatjanski and Levenštein [1]).

$$\delta(S_n) \le 2^{-0.599n(1+o(1))}.$$

Proof. Routine computation yields

$$\frac{1 + \sin \varphi}{2 \sin \varphi} \log \frac{1 + \sin \varphi}{2 \sin \varphi} - \frac{1 - \sin \varphi}{2 \sin \varphi} \log \frac{1 - \sin \varphi}{2 \sin \varphi}$$

$$\le -\frac{1}{2} \log(1 - \cos \varphi) - 0.099$$

when $0 < \varphi \le \varphi_0 \approx 63°$. Then, it follows from Lemma 8.2 that

$$m[n, \varphi] \le (1 - \cos \varphi)^{-0.5n} 2^{-0.099n}$$

$$= \left(\sin \frac{\varphi}{2} \right)^{-n} 2^{-0.599n}. \qquad (8.21)$$

Consequently, by Lemma 8.3 we have

$$\delta(S_n) \le \left(\sin \frac{\varphi}{2} \right)^n \left(\sin \frac{\varphi}{2} \right)^{-(n+1)} 2^{-0.599(n+1)}$$

$$= \left(\sin \frac{\varphi}{2} \right)^{-1} 2^{-0.599n(1+o(1))}$$

for some suitable φ. Theorem 8.2 follows. □

To end this chapter let us propose two open problems.

Problem 8.1. *What is the behavior of*

$$f(n) = k(S_n) - k^*(S_n)?$$

Does $f(n) = 0$ (or $\ne 0$) hold for infinitely many dimensions? What is the asymptotic order of $\overline{\lim} \, f(n)$?

Problem 8.2. *What is the behavior of*

$$g(n) = \frac{\delta(S_n)}{\delta^*(S_n)}?$$

Does $g(n) = 1$ (or $\ne 1$) hold for infinitely many dimensions? What is the asymptotic order of $\overline{\lim} \, g(n)$?

9. The Kissing Numbers of Spheres in Eight and Twenty–Four Dimensions

9.1. Some Special Lattices

Let Λ be an n-dimensional lattice. As usual, we call it an *integral lattice* if

$$\langle \mathbf{u}, \mathbf{u} \rangle \in Z, \quad \mathbf{u} \in \Lambda.$$

In particular, we call it an *even integral lattice* if

$$\langle \mathbf{u}, \mathbf{u} \rangle \in 2Z, \quad \mathbf{u} \in \Lambda,$$

and an *odd integral lattice* otherwise.

There are some special integral lattices that play important roles in different areas of mathematics. For example,

$$Z_n = \left\{ \mathbf{z} \in E^n : z_i \in Z \right\},$$

$$A_n := \left\{ \mathbf{z} \in Z_{n+1} : \sum_{i=1}^{n+1} z_i = 0 \right\},$$

$$D_n = \left\{ \mathbf{z} \in Z_n : \sum_{i=1}^{n} z_i \in 2Z \right\},$$

$$E_8 = \left\{ \mathbf{u} \in \frac{1}{2} Z_8 : u_i - u_j \in Z, \ \sum_{i=1}^{8} u_i \in 2Z \right\},$$

$$E_7 = \left\{ \mathbf{u} \in E_8 : \sum_{i=1}^{8} u_i = 0 \right\},$$

$$E_6 = \left\{ \mathbf{u} \in E_8 : \sum_{i=1}^{6} u_i = u_7 + u_8 = 0 \right\},$$

and the well-known Leech lattice Λ_{24}, which is determined by the basis $\mathbf{a}_1 = 2\sqrt{2}\mathbf{e}_1$, $\mathbf{a}_2 = \sqrt{2}(\mathbf{e}_1 + \mathbf{e}_2)$, $\mathbf{a}_3 = \sqrt{2}(\mathbf{e}_1 + \mathbf{e}_3)$, $\mathbf{a}_4 = \sqrt{2}(\mathbf{e}_1 + \mathbf{e}_4)$, $\mathbf{a}_5 = \sqrt{2}(\mathbf{e}_1 + \mathbf{e}_5)$, $\mathbf{a}_6 = \sqrt{2}(\mathbf{e}_1 + \mathbf{e}_6)$, $\mathbf{a}_7 = \sqrt{2}(\mathbf{e}_1 + \mathbf{e}_7)$, $\mathbf{a}_8 = (\mathbf{e}_1 + \cdots + \mathbf{e}_8)/\sqrt{2}$, $\mathbf{a}_9 = \sqrt{2}(\mathbf{e}_1 + \mathbf{e}_9)$, $\mathbf{a}_{10} = \sqrt{2}(\mathbf{e}_1 + \mathbf{e}_{10})$, $\mathbf{a}_{11} = \sqrt{2}(\mathbf{e}_1 + \mathbf{e}_{11})$, $\mathbf{a}_{12} = (\mathbf{e}_1 + \cdots + \mathbf{e}_4 + \mathbf{e}_9 + \cdots + \mathbf{e}_{12})/\sqrt{2}$, $\mathbf{a}_{13} = \sqrt{2}(\mathbf{e}_1 + \mathbf{e}_{13})$, $\mathbf{a}_{14} = (\mathbf{e}_1 + \mathbf{e}_2 + \mathbf{e}_5 + \mathbf{e}_6 + \mathbf{e}_9 + \mathbf{e}_{10} + \mathbf{e}_{13} + \mathbf{e}_{14})/\sqrt{2}$, $\mathbf{a}_{15} = (\mathbf{e}_1 + \mathbf{e}_3 + \mathbf{e}_5 + \mathbf{e}_7 + \mathbf{e}_9 + \mathbf{e}_{11} + \mathbf{e}_{13} + \mathbf{e}_{15})/\sqrt{2}$, $\mathbf{a}_{16} = (\mathbf{e}_1 + \mathbf{e}_4 + \mathbf{e}_5 + \mathbf{e}_8 + \mathbf{e}_9 + \mathbf{e}_{12} + \mathbf{e}_{13} + \mathbf{e}_{16})/\sqrt{2}$, $\mathbf{a}_{17} = \sqrt{2}(\mathbf{e}_1 + \mathbf{e}_{17})$, $\mathbf{a}_{18} = (\mathbf{e}_1 + \mathbf{e}_3 + \mathbf{e}_5 + \mathbf{e}_8 + \mathbf{e}_9 + \mathbf{e}_{10} + \mathbf{e}_{17} + \mathbf{e}_{18})/\sqrt{2}$, $\mathbf{a}_{19} = (\mathbf{e}_1 + \mathbf{e}_4 + \mathbf{e}_5 + \mathbf{e}_6 + \mathbf{e}_9 + \mathbf{e}_{11} + \mathbf{e}_{17} + \mathbf{e}_{19})/\sqrt{2}$, $\mathbf{a}_{20} = (\mathbf{e}_1 + \mathbf{e}_2 + \mathbf{e}_5 + \mathbf{e}_7 + \mathbf{e}_9 + \mathbf{e}_{12} + \mathbf{e}_{17} + \mathbf{e}_{20})/\sqrt{2}$, $\mathbf{a}_{21} = (\mathbf{e}_2 + \mathbf{e}_3 + \mathbf{e}_4 + \mathbf{e}_5 + \mathbf{e}_9 + \mathbf{e}_{13} + \mathbf{e}_{17} + \mathbf{e}_{21})/\sqrt{2}$, $\mathbf{a}_{22} = (\mathbf{e}_9 + \mathbf{e}_{10} + \mathbf{e}_{13} + \mathbf{e}_{14} + \mathbf{e}_{17} + \mathbf{e}_{18} + \mathbf{e}_{21} + \mathbf{e}_{22})/\sqrt{2}$, $\mathbf{a}_{23} = (\mathbf{e}_9 + \mathbf{e}_{11} + \mathbf{e}_{13} + \mathbf{e}_{15} + \mathbf{e}_{17} + \mathbf{e}_{19} + \mathbf{e}_{21} + \mathbf{e}_{23})/\sqrt{2}$, $\mathbf{a}_{24} = (-3\mathbf{e}_1 + \mathbf{e}_2 + \cdots + \mathbf{e}_{24})/\sqrt{8}$, where $\mathbf{e}_1, \mathbf{e}_2, \ldots, \mathbf{e}_{24}$ is an orthonormal basis of E^{24}. It can be verified that A_n $(n \geq 1)$, D_n $(n \geq 4)$, E_n $(n = 6, 7, 8)$, and Λ_{24} are even integral lattices, and Z_n $(n \geq 1)$ are odd integral lattices.

Writing

$$r = \min \left\{ \tfrac{1}{2}\|\mathbf{u}\| : \mathbf{u} \in \Lambda \setminus \{\mathbf{o}\} \right\}$$

and

$$M(\Lambda) = \{\mathbf{u} \in \Lambda : \|\mathbf{u}\| = 2r\},$$

it is easy to see that $rS_n + \Lambda$ is a packing and $rS_n + \mathbf{u}$ touches rS_n at its boundary if $\mathbf{u} \in M(\Lambda)$. Thus,

$$k^*(S_n) = \max \{\text{card}\{M(\Lambda)\}\}, \tag{9.1}$$

where the maximum is over all n-dimensional lattices. In particular, routine computation yields

$$\text{card}\{M(E_8)\} = 240 \tag{9.2}$$

and

$$\text{card}\{M(\Lambda_{24})\} = 196560. \tag{9.3}$$

To end this section we cite a well-known result about the representations of integral lattices that will be useful in Section 4 of this chapter. We omit its extremely complicated proof.

Lemma 9.1 (Kneser [1]). *Every integral lattice generated by vectors of norm 1 and $\sqrt{2}$ can be written as a direct sum of the special lattices Z_n $(n \geq 1)$, A_n $(n \geq 1)$, D_n $(n \geq 4)$, and E_n $(n = 6, 7, 8)$.*

9.2. Two Theorems of Levenštein, Odlyzko, and Sloane

Theorem 9.1 (Levenštein [2], Odlyzko and Sloane [1]).

$$k^*(S_8) = k(S_8) = 240.$$

Proof. Abbreviate the Jacobi polynomial $P_k^{2.5,2.5}(t)$ to P_k, and define

$$f(t) = \frac{320}{3}(t+1)\left(t+\frac{1}{2}\right)^2 t^2 \left(t-\frac{1}{2}\right)$$
$$= P_0 + \frac{16}{7}P_1 + \frac{200}{63}P_2 + \frac{832}{231}P_3$$
$$+ \frac{1216}{429}P_4 + \frac{5120}{3003}P_5 + \frac{2560}{4641}P_6.$$

Clearly, $f(t)$ satisfies the conditions of Lemma 8.1 for $\varphi = \pi/3$. Thus, by (8.1) and Lemma 8.1 it follows that

$$k(S_8) = m(8, \pi/6) \leq \frac{f(1)}{f_0} = 240. \qquad (9.4)$$

Then, (9.1), (9.2), and (9.4) together yield

$$k^*(S_8) = k(S_8) = 240.$$

Theorem 9.1 is proved.   □

Theorem 9.2 (Levenštein [2], Odlyzko and Sloane [1]).

$$k^*(S_{24}) = k(S_{24}) = 196560.$$

Proof. Abbreviate the Jacobi polynomial $P_k^{10.5,10.5}(t)$ to P_k, and define

$$f(t) = \frac{1490944}{15}(t+1)\left(t+\frac{1}{2}\right)^2\left(t+\frac{1}{4}\right)^2 t^2 \left(t-\frac{1}{4}\right)^2 \left(t-\frac{1}{2}\right)$$
$$= P_0 + \frac{48}{23}P_1 + \frac{1144}{425}P_2 + \frac{12992}{3825}P_3 + \frac{73888}{22185}P_4$$
$$+ \frac{2169856}{687735}P_5 + \frac{59062016}{25365285}P_6 + \frac{4472832}{2753575}P_7$$
$$+ \frac{23855104}{28956015}P_8 + \frac{7340032}{20376455}P_9 + \frac{7340032}{80848515}P_{10}.$$

Then, $f(t)$ satisfies the conditions of Lemma 8.1 for $\varphi = \pi/3$. Thus, by (8.1) and Lemma 8.1 we have

$$k(S_{24}) = m(24, \pi/6) \leq \frac{f(1)}{f_0} = 196560. \qquad (9.5)$$

Then, (9.1), (9.3), and (9.5) together yield

$$k^*(S_{24}) = k(S_{24}) = 196560.$$

Theorem 9.2 is proved.   □

Remark 9.1. *By constructing a suitable polynomial one can prove that*

$$k(S_4) \leq 25.$$

On the other hand, by Theorem 2.6, we know that

$$k^*(S_4) = 24.$$

Thus, $k(S_4)$ is either 24 or 25. However, the correct answer is still unknown.

Remark 9.2. *By choosing suitable polynomials, numerical upper bounds for $\delta(S_n)$ and $k(S_n)$ can be obtained in special dimensions. For results of this kind we refer to Conway and Sloane [1].*

9.3. Two Principles of Linear Programming

For convenience, for two points **u** and **v** of E^n we write $\mathbf{u} \preceq \mathbf{v}$ if $u_i \leq v_i$ holds simultaneously for $i = 1, 2, \ldots, n$. Usually, a linear programming problem is to determine

$$p = \max \left\{ \langle \mathbf{a}, \mathbf{x} \rangle : \ \mathbf{x}A' \preceq \mathbf{b}, \ \mathbf{o} \preceq \mathbf{x} \right\}, \tag{9.6}$$

where $\mathbf{a} \in E^n$, $\mathbf{b} \in E^m$, and A is a $m \times n$ matrix. Its dual is to determine

$$q = \min \left\{ \langle \mathbf{b}, \mathbf{y} \rangle : \ \mathbf{a} \preceq \mathbf{y}A, \ \mathbf{o}_1 \preceq \mathbf{y} \right\}, \tag{9.7}$$

where $\mathbf{o}_1$ is the origin of E^m.

Now we introduce two fundamental principles concerning these problems that will be useful in the next section. Since they are well-known and can be found in any book on linear programming (see Grötschel, Lovász, and Schrijver [1] or Nemhauser and Wolsey [1]), we omit their proofs here.

Principle 9.1. *If p or q is finite, then $p = q$.*

Principle 9.2. *If $\mathbf{x}^*$ is an optimal solution of (9.6) and $\mathbf{y}^*$ is an optimal solution of (9.7), writing $\mathbf{s}^* = \mathbf{b} - \mathbf{x}^*A'$ and $\mathbf{t}^* = \mathbf{y}^*A - \mathbf{a}$, then*

$$x_i^* t_i^* = 0, \quad i = 1, 2, \ldots, n,$$

and

$$y_j^* s_j^* = 0, \quad j = 1, 2, \ldots, m.$$

Remark 9.3. *In fact, Principle 9.2 can be easily deduced from Principle 9.1.*

9.4. Two Theorems of Bannai and Sloane

Let $\aleph$ be an $\{n, m, \varphi\}$ code. For $\mathbf{u} \in \aleph$ the *inner product distribution* of $\aleph$ with respect to $\mathbf{u}$ is the set of numbers $\{a(\mathbf{u}, t) : -1 \leq t \leq 1\}$, where

$$a(\mathbf{u}, t) = \operatorname{card} \{\mathbf{v} \in \aleph : \langle \mathbf{u}, \mathbf{v} \rangle = t\},$$

and the inner product distribution of $\aleph$ is the set of numbers $\{a(t) : -1 \leq t \leq 1\}$, where

$$a(t) = \frac{1}{m} \sum_{\mathbf{u} \in \aleph} a(\mathbf{u}, t).$$

In particular, writing $\gamma = \cos \varphi$, we have $a(1) = 1$, $a(t) = 0$ for $\gamma < t < 1$, and

$$m - 1 = \sum_{-1 \leq t \leq \gamma} a(t).$$

Also, by Assertions 8.10 and 8.9, when $\alpha = (n-3)/2$ we have

$$\sum_{-1 \leq t \leq 1} a(t) P_k^{\alpha,\alpha}(t) = \frac{1}{mc_k} \sum_{\mathbf{x}, \mathbf{y} \in \aleph} Q_k(\mathbf{x}, \mathbf{y})$$

$$= \frac{1}{mc_k} \sum_{\mathbf{x}, \mathbf{y} \in \aleph} \sum_{i=1}^{d(k)} h_{k,i}(\mathbf{x}) h_{k,i}(\mathbf{y})$$

$$= \frac{1}{mc_k} \sum_{i=1}^{d(k)} \left(\sum_{\mathbf{x} \in \aleph} h_{k,i}(\mathbf{x}) \right)^2 \geq 0. \qquad (9.8)$$

Thus, for fixed n and φ an upper bound to m is given by the following linear programming problem:

$$p(n, \varphi) = \max \left\{ 1 + \sum_{-1 \leq t \leq \gamma} a(t) : -\sum_{-1 \leq t \leq \gamma} a(t) P_k^{\alpha,\alpha}(t) \leq P_k^{\alpha,\alpha}(1) \right.$$

$$\left. \text{for } k \geq 1, \ a(t) \geq 0 \right\}. \qquad (9.9)$$

Its dual can be formulated as

$$q(n,\varphi) = \min\left\{\frac{f(1)}{f_0} : f(t) = \sum_{k=0}^{l} f_k P_k^{\alpha,\alpha}(t), \ f_0 > 0, \ f_k \geq 0 \text{ for } k \geq 1;\right.$$

$$\left. f(t) \leq 0 \text{ for } -1 \leq t \leq \gamma \right\}. \tag{9.10}$$

Remark 9.4. *Formulae* (9.9), (9.10), *and Principle* 9.1 *together imply Lemma* 8.1.

Theorem 9.3 (Bannai and Sloane [1]). *There is a unique way (up to isometry) of arranging* 240 *nonoverlapping unit spheres in* E^8 *so that they all touch another unit sphere.*

Proof. Let $\aleph^* = \{u_1^*, u_2^*, \ldots, u_{240}^*\}$ be an $\{8, 240, \pi/3\}$ code, and let $\{a^*(t) : -1 \leq t \leq 1\}$ be its inner product distribution. Then $\{a^*(t) : -1 \leq t \leq 1\}$ is an optimal solution to the linear programming problem (9.9) for $n = 8$ and $\varphi = \pi/3$, and the polynomial $f(t)$ defined in the proof of Theorem 9.1 is an optimal solution to its dual problem (9.10). Since the dual variables $f_1, f_2, \ldots, f_6$ are nonzero, by Principle 9.2, (9.8) holds with equality for $k = 1, 2, \ldots, 6$. On the other hand, since the optimal polynomial $f(t)$ does not vanish, except for $t = -1, -\frac{1}{2}, 0,$ and $\frac{1}{2}$, the primal variables must vanish everywhere except perhaps for $a^*(-1)$, $a^*(-\frac{1}{2})$, $a^*(0)$, and $a^*(\frac{1}{2})$. From (9.8), these numbers satisfy the equations

$$\sum_{-1 \leq t < 1} a^*(t) P_k^{\alpha,\alpha}(t) = -P_k^{\alpha,\alpha}(1)$$

for $k = 1, 2, \ldots, 6$. Solving these equations we obtain

$$a^*(-1) = 1, \quad a^*(-\tfrac{1}{2}) = a^*(\tfrac{1}{2}) = 56, \quad \text{and} \quad a^*(0) = 126,$$

which implies that the possible inner products in $\aleph^*$ are $0, \pm\frac{1}{2},$ and ± 1. Thus, defining

$$\Lambda^* = \left\{ \sum_{i=1}^{240} \sqrt{2} z_i u_i^* : z_i \in Z \right\},$$

it is easy to verify that Λ^* is an even integral lattice, and

$$M(\Lambda^*) = \sqrt{2}\,\aleph^*.$$

Then it follows from Lemma 9.1 that Λ^* is a direct sum of the lattices A_n $(n \geq 1)$, D_n $(n \geq 4)$, and E_n $(n = 6, 7, 8)$. In fact, the only lattice of this type with at least 240 minimal vectors is E_8. Thus, Λ^* is isometric to E_8, and therefore $\sqrt{2}\aleph^*$ is isometric to $M(E_8)$. Theorem 9.3 is proved. $\square$

Theorem 9.4 (Bannai and Sloane [1]). *There is a unique way (up to isometry) of arranging* 196560 *nonoverlapping unit spheres in* E^{24} *so that they all touch another unit sphere.*

Let $\aleph^* = \{\mathbf{u}_1^*, \mathbf{u}_2^*, \ldots, \mathbf{u}_{196560}^*\}$ be a $\{24, 196560, \pi/3\}$ code. We proceed to show that $2\aleph^*$ is isometric to $M(\Lambda_{24})$. In other words, writing

$$\Lambda^* = \left\{ \sum_{i=1}^{196560} 2z_i\mathbf{u}_i^* : z_i \in Z \right\},$$

we try to prove that $2\aleph^* = M(\Lambda^*)$ and Λ^* is isometric to Λ_{24}. This requires some further preparation.

A spherical code $\aleph$ of dimension n is called a *spherical l-design* if

$$\sum_{\mathbf{x}\in\aleph} h_k(\sigma\mathbf{x}) = \sum_{\mathbf{x}\in\aleph} h_k(\mathbf{x}) \qquad (9.11)$$

holds for every homogeneous polynomial h_k of degree $k \leq l$ and for every isometry σ of E^n. It has the following defining property.

Lemma 9.2 (Delsarte, Goethals, and Seidel [1]). *Let $\aleph$ be a spherical code of dimension n, cardinality m, and inner product distribution $\{a(t) : -1 \leq t \leq 1\}$. Then $\aleph$ is a spherical l-design if and only if*

$$\sum_{-1\leq t\leq 1} a(t)P_k^{\alpha,\alpha}(t) = 0$$

holds for $\alpha = (n-3)/2$ and $k = 1, 2, \ldots, l$.

Proof. It follows from (9.11) that $\aleph$ is a spherical l-design if

$$\frac{1}{m}\sum_{\mathbf{x}\in\aleph} h_k(\sigma\mathbf{x}) = \frac{1}{n\omega_n}\int_{\mathrm{bd}(S_n)} h_k(\mathbf{x})d\omega(\mathbf{x}) \qquad (9.12)$$

holds for every homogeneous polynomial $h_k(\mathbf{x})$ of degree $k \leq l$ and for every isometry σ of E^n. It is well-known that

$$\int_{\mathrm{bd}(S_n)} h_k(\mathbf{x})d\omega(\mathbf{x}) = 0$$

holds for every spherical harmonic polynomial $h_k(\mathbf{x})$ of degree $k \geq 1$. Thus, by (9.12), (8.5), and Assertion 8.7, $\aleph$ is a spherical l-design if and only if

$$\sum_{\mathbf{x}\in\aleph} h_k(\mathbf{x}) = 0 \qquad (9.13)$$

holds for every spherical harmonic polynomial of degree k, $1 \leq k \leq l$.

Let $h_{k,1}, h_{k,2}, \ldots, h_{k,d(k)}$ be a basis for $\mathcal{H}_k$. By Assertions 8.9 and 8.10, we have

$$P_k^{\alpha,\alpha}(\langle\mathbf{x},\mathbf{y}\rangle) = \frac{1}{c_k}\sum_{i=1}^{d(k)} h_{k,i}(\mathbf{x})h_{k,i}(\mathbf{y})$$

for some positive number c_k. Hence

$$\sum_{-1 \leq t \leq 1} a(t) P_k^{\alpha,\alpha}(t) = \frac{1}{m} \sum_{\mathbf{x}, \mathbf{y} \in \aleph} P_k^{\alpha,\alpha}(\langle \mathbf{x}, \mathbf{y} \rangle)$$

$$= \frac{1}{mc_k} \sum_{i=1}^{d(k)} \left(\sum_{\mathbf{x} \in \aleph} h_{k,i}(\mathbf{x}) \right)^2.$$

Thus, by (9.13), $\aleph$ is a spherical l-design if and only if

$$\sum_{-1 \leq t \leq 1} a(t) P_k^{\alpha,\alpha}(t) = 0$$

holds for $1 \leq k \leq l$. Lemma 9.2 is proved. $\square$

Corollary 9.1. $\frac{1}{2} M(\Lambda_{24})$ *is a spherical 11-design.*

Lemma 9.3 (Bannai and Sloane [1]). *The code $\aleph^\star$ is a spherical 11-design. Further, letting $\{a^\star(t) : -1 \leq t \leq 1\}$ be its inner product distribution,*

$$a^\star(t) = \begin{cases} 1 & \text{if } t = \pm 1, \\ 4600 & \text{if } t = \pm\frac{1}{2}, \\ 47104 & \text{if } t = \pm\frac{1}{4}, \\ 93150 & \text{if } t = 0, \\ 0 & \text{otherwise.} \end{cases}$$

This lemma can be proved by a similar argument to that of the first part of the proof of Theorem 9.3.

Lemma 9.4 (Bannai and Sloane [1]). $\Lambda^\star$ *is an even integral lattice such that*

$$\min \{\langle \mathbf{v}, \mathbf{v} \rangle : \ \mathbf{v} \in \Lambda^\star \setminus \{\mathbf{o}\}\} = 4.$$

Proof. By Lemma 9.3 it is easy to see that $\Lambda^\star$ is an even integral lattice. Suppose that $\langle \mathbf{v}, \mathbf{v} \rangle = 2$ holds for some point $\mathbf{v} \in \Lambda^\star$, say $\mathbf{v} = \sqrt{2}\mathbf{e}_1$. Then

$$\langle \mathbf{u}, \mathbf{v} \rangle \in \{0, \pm 1, \pm 2\}, \quad \mathbf{u} \in 2\aleph^\star,$$

since $2\langle \mathbf{u}, \mathbf{v} \rangle = \langle \mathbf{u} + \mathbf{v}, \mathbf{u} + \mathbf{v} \rangle - \langle \mathbf{u}, \mathbf{u} \rangle - \langle \mathbf{v}, \mathbf{v} \rangle$ is an even integer and $|\langle \mathbf{u}, \mathbf{v} \rangle| \leq 2\sqrt{2}$. In other words,

$$\langle \mathbf{u}_i^\star, \mathbf{v} \rangle \in \left\{0, \pm\tfrac{1}{2}, \pm 1\right\}, \quad \mathbf{u}_i^\star \in \aleph^\star.$$

Now we proceed to deduce a contradiction.

Suppose $\langle \mathbf{u}_i^\star, \mathbf{v} \rangle = 0$ for m_1 choices of $\mathbf{u}_i^\star$, $|\langle \mathbf{u}_i^\star, \mathbf{v} \rangle| = \frac{1}{2}$ for m_2 choices, and $|\langle \mathbf{u}_i^\star, \mathbf{v} \rangle| = 1$ for m_3 choices. Clearly,

$$m_1 + m_2 + m_3 = 196560.$$

Since $\aleph^\star$ is a spherical 11-design, then by (9.12),

$$\frac{1}{196560} \sum_{\mathbf{u}_i^\star \in \aleph^\star} h_k(\mathbf{u}_i^\star) = \frac{1}{24\omega_{24}} \int_{\mathrm{bd}(S_{24})} h_k(\mathbf{x}) d\omega(\mathbf{x}) \qquad (9.14)$$

holds for any homogeneous polynomial $h_k(\mathbf{x})$ of degree $k \le 11$. Let us choose $h_k(\mathbf{x}) = (x_1)^k$, for $k = 2$ and 4, such that

$$h_k(\mathbf{u}_i^\star) = \left(\frac{\langle \mathbf{u}_i^\star, \mathbf{v} \rangle}{\sqrt{2}} \right)^k.$$

By (9.12) and Corollary 9.1, the right-hand side of (9.14) can be evaluated by

$$\frac{1}{24\omega_{24}} \int_{\mathrm{bd}(S_{24})} h_k(\mathbf{x}) d\omega(\mathbf{x}) = \frac{1}{196560} \sum_{\mathbf{u} \in \frac{1}{2} M(\Lambda_{24})} h_k(\mathbf{u})$$

$$= \begin{cases} \frac{8190}{196560} & \text{if } k = 2, \\ \frac{945}{196560} & \text{if } k = 4. \end{cases}$$

Then, taking $k = 2$ and 4, respectively, it follows from (9.14) that

$$\begin{cases} \frac{1}{8}m_2 + \frac{1}{2}m_3 = 8190, \\ \frac{1}{64}m_2 + \frac{1}{4}m_3 = 945, \end{cases}$$

which implies $m_2 = 67200$ and $m_3 = -420$, an impossibility. Thus, Lemma 9.4 is proved. □

Corollary 9.2.

$$2\aleph^\star = M(\Lambda^\star).$$

Lemma 9.5 (Bannai and Sloane [1]). *Let $\aleph^\star$ be a $\{24, 196560, \pi/3\}$ code. For any pair of vectors $\mathbf{u}^\star$ and $\mathbf{v}^\star$ in $\aleph^\star$ with $\langle \mathbf{u}^\star, \mathbf{v}^\star \rangle = 0$ there are 44 vectors $\mathbf{w}^\star \in \aleph^\star$ such that*

$$\langle \mathbf{u}^\star, \mathbf{w}^\star \rangle = \langle \mathbf{v}^\star, \mathbf{w}^\star \rangle = \tfrac{1}{2}. \qquad (9.15)$$

By Lemma 9.3, the code $\aleph^\star$ is an 11-design. Define

$$h_k(\mathbf{x}) = f(\mathbf{x})^k = (\langle \mathbf{u}^\star, \mathbf{x} \rangle + \langle \mathbf{v}^\star, \mathbf{x} \rangle)^k.$$

Clearly, $h_k(\mathbf{x})$ is a homogeneous polynomial. Also, by Lemma 9.3, $f(\mathbf{w}^\star)$ takes only the eight different nonzero values ± 1, $\pm \frac{3}{4}$, $\pm \frac{1}{2}$, and $\pm \frac{1}{4}$ for $\mathbf{w}^\star \in \aleph^\star$, and the number of points $\mathbf{w}^\star \in \aleph^\star \setminus \{\mathbf{u}^\star, \mathbf{v}^\star\}$ such that $f(\mathbf{w}^\star) = 1$ is exactly the number of points satisfying (9.15). Considering the cases

$k = 1, 2, \ldots, 8$, this lemma can be proved by a similar argument to that of the proof of Lemma 9.4.

Geometrically speaking, Lemma 9.5 means that for any pair of vectors $\mathbf{u}^\star$ and $\mathbf{v}^\star$ in $\aleph^\star$ with $\angle\mathbf{u}^\star o \mathbf{v}^\star = \pi/2$ there are 44 vectors $\mathbf{w}^\star \in \aleph^\star$ such that

$$\angle\mathbf{u}^\star o \mathbf{w}^\star = \angle\mathbf{v}^\star o \mathbf{w}^\star = \pi/3.$$

For $n \geq 3$ we define

$$G_n = \left\{ \sum_{i=1}^{n} z_i \mathbf{w}_i : z_i \in Z \right\},$$

where $\mathbf{w}_1 = \sqrt{2}(\mathbf{e}_1 + \mathbf{e}_2)$ and $\mathbf{w}_i = \sqrt{2}(\mathbf{e}_{i-1} - \mathbf{e}_i)$ for $i = 2, 3, \ldots, n$. In fact, $G_n = \sqrt{2}D_n$. It is easy to check that $\angle\mathbf{w}_1 o \mathbf{w}_2 = \pi/2$, $\angle\mathbf{w}_i o \mathbf{w}_{i+1} = \pi/3$ $(i = 2, 3, \ldots, n-1)$, $G_3 \subset G_4 \subset \cdots$, and there are $2n(n-1)$ minimal vectors of $[\sqrt{2}^2 | 0^{n-2}]$ type in G_n.

Lemma 9.6 (Bannai and Sloane [1]). *For $n = 3, 4, \ldots, 24$, the lattice $\Lambda^\star$ contains a suitable sublattice isometric to G_n.*

Proof. It follows from Lemma 9.5 that $\Lambda^\star$ contains a suitable sublattice isometric to G_3. Now we proceed by induction on n. Suppose the assertion holds for $n = k$, $3 \leq k < 24$. By choosing a suitable orthonormal basis $\mathbf{e}_1$, $\mathbf{e}_2, \ldots, \mathbf{e}_{24}$, $M(\Lambda^\star)$ contains vectors $\mathbf{w}_1, \mathbf{w}_2, \ldots, \mathbf{w}_k$ that span G_k. By Lemma 9.5 there are 44 vectors $\mathbf{w} \in M(\Lambda^\star)$ with $\angle\mathbf{w}_1 o \mathbf{w} = \angle\mathbf{w}_2 o \mathbf{w} = \pi/3$. On the other hand, it can be verified that G_k has $2k - 4 < 44$ minimal vectors $\mathbf{w}'$ such that $\angle\mathbf{w}_1 o \mathbf{w}' = \angle\mathbf{w}_2 o \mathbf{w}' = \pi/3$. Thus, at least one of the 44 vectors $\mathbf{w}$, say $\mathbf{w}_{k+1}$, is not a minimal vector of G_k.

Now we proceed to show that $\mathbf{w}_{k+1} \notin E^k$, where E^k indicates the space spanned by $\mathbf{w}_1, \mathbf{w}_2, \ldots, \mathbf{w}_k$. Suppose that, on the contrary,

$$\mathbf{w}_{k+1} = w_1 \mathbf{e}_1 + w_2 \mathbf{e}_2 + \cdots + w_k \mathbf{e}_k.$$

By $\angle\mathbf{w}_1 o \mathbf{w}_{k+1} = \angle\mathbf{w}_2 o \mathbf{w}_{k+1} = \pi/3$ it follows that $w_1 = \sqrt{2}$ and $w_2 = 0$. For $3 \leq i \leq k$,

$$\sqrt{2}(\mathbf{e}_1 \pm \mathbf{e}_i) \in M(\Lambda^\star) \cap G_k \subseteq 2\aleph^\star$$

and therefore

$$\langle \mathbf{w}_{k+1}, \sqrt{2}(\mathbf{e}_1 \pm \mathbf{e}_i) \rangle \in \{0, \pm1, \pm2\},$$

which implies $w_3 = w_4 = \cdots = w_k = 0$. Then, we obtain $\langle \mathbf{w}_{k+1}, \mathbf{w}_{k+1} \rangle = 2$, which contradicts $\mathbf{w}_{k+1} \in 2\aleph^\star$. Thus, we have $\mathbf{w}_{k+1} \notin E^k$.

Choose $\mathbf{e}_{k+1}$ such that $\{\mathbf{e}_1, \mathbf{e}_2, \ldots, \mathbf{e}_{k+1}\}$ is an orthonormal basis for the $(k+1)$-dimensional space E^{k+1} that contains E^k and $\mathbf{w}_{k+1}$, and suppose that

$$\mathbf{w}_{k+1} = w_1 \mathbf{e}_1 + w_2 \mathbf{e}_2 + \cdots + w_{k+1} \mathbf{e}_{k+1}.$$

The previous argument shows that $w_1 = \sqrt{2}$, $w_2 = w_3 = \cdots = w_k = 0$, and $w_{k+1} = \pm\sqrt{2}$. Hence,

$$G_{k+1} = \left\{ \sum_{i=1}^{k+1} z_i \mathbf{w}_i : z_i \in Z \right\} \subseteq \Lambda^\star.$$

Lemma 9.6 is proved.   □

Proof of Theorem 9.4. By Lemma 9.6 we may choose an orthonormal basis e_1, e_2, ..., e_{24} in E^{24} such that $G_{24} \subset \Lambda^\star$, and therefore $M(\Lambda^\star)$ contains all the points $\mathbf{v}^\star$ of $[4^2|0^{22}]/\sqrt{8}$ type. Let $\mathbf{u}^\star = (u_1^\star, u_2^\star, \ldots, u_{24}^\star)/\sqrt{8}$ be any point of $M(\Lambda^\star)$. Then

$$\sum_{i=1}^{24} (u_i^\star)^2 = 32. \tag{9.16}$$

On the other hand, it follows from Lemma 9.3 that

$$\langle \mathbf{u}^\star, \mathbf{v}^\star \rangle = \{0, \pm 1, \pm 2, \pm 4\}.$$

In other words,

$$\tfrac{1}{2}(u_i^\star \pm u_j^\star) \in \{0, \pm 1, \pm 2 \pm 4\} \tag{9.17}$$

for all indices $i \neq j$. By (9.17) and (9.16) we have

$$u_i^\star = \tfrac{1}{2}(u_i^\star + u_j^\star) + \tfrac{1}{2}(u_i^\star - u_j^\star) \in \{0, \pm 1, \pm 2, \pm 3, \pm 4, \pm 5\}.$$

Also, it follows from (9.17) that

$$u_1^\star \equiv u_2^\star \equiv \cdots \equiv u_{24}^\star \pmod{2}$$

and, if $|u_i^\star| = 4$ for one index i,

$$u_1^\star \equiv u_2^\star \equiv \cdots \equiv u_{24}^\star \pmod{4}. \tag{9.18}$$

Thus, as a conclusion, the only possibilities for the components of $\mathbf{u}^\star \in M(\Lambda^\star)$ are

$$[2^8|0^{16}]/\sqrt{8}, \quad [4^2|0^{22}]/\sqrt{8}, \quad \text{and} \quad [3^1|1^{23}]/\sqrt{8}.$$

Denote by M_1, M_2, and M_3 respectively the subsets of $M(\Lambda^\star)$ consisting of points of the above types. Now we study the structure of these subsets.

Case 1. Let $F(\mathbf{x}) = (f(x_1), f(x_2), \ldots, f(x_{24}))$ be a map defined by

$$f(x) = \begin{cases} 0 & \text{if } x = 0, \\ 1 & \text{otherwise,} \end{cases}$$

and define

$$\aleph_1 = \{F(\mathbf{u}^*) : \ \mathbf{u}^* \in M_1\}.$$

Clearly, the weight of every codeword of $\aleph_1$ is eight. If $F(\mathbf{u}^*) \neq F(\mathbf{v}^*)$ and $f(u_i^*) = f(v_i^*) = 1$ for at least five indices for two points $\mathbf{u}^*$ and $\mathbf{v}^*$ of M_1, then by (9.18) one can find a point $\mathbf{w}^* \in G_{24}$ such that $\|\mathbf{u}^* + \mathbf{v}^* + \mathbf{w}^*\| < 2$. This is impossible. Thus,

$$\text{card}\{\aleph_1\} \leq \binom{24}{5} \bigg/ \binom{8}{5} = 759.$$

Also, without loss of generality, if $\mathbf{u}^* = (u_1^*, u_2^*, \ldots, u_{24}^*)/\sqrt{8} \in M_1$ and $u_1^* \neq 0$, then $(-u_i^*, u_2^*, \ldots, u_{24}^*)/\sqrt{8} \notin M_1$. This means that at most 2^7 different points of M_1 correspond to one codeword of $\aleph_1$. Hence,

$$\text{card}\{M_1\} \leq 759 \times 2^7 = 97152. \tag{9.19}$$

Case 2. Clearly,

$$\text{card}\{M_2\} = \binom{24}{2} 2^2 = 1104. \tag{9.20}$$

Case 3. Let $G(\mathbf{x}) = (g(x_1), g(x_2), \ldots, g(x_{24}))$ be a map defined by

$$g(x) = \begin{cases} 1 & \text{if } x = 1/\sqrt{8}, \\ 1 & \text{if } x = -3/\sqrt{8}, \\ 0 & \text{otherwise,} \end{cases}$$

and define

$$\aleph_3 = \{G(\mathbf{u}^*) : \ \mathbf{u}^* \in M_3\}.$$

Then, routine arguments yield

$$\|\mathbf{u}, \mathbf{v}\|_H \geq 8$$

for any pair of distinct codewords $\mathbf{u}$ and $\mathbf{v}$ of $\aleph_3$. We now proceed to estimate $\text{card}\{\aleph_3\}$ and $\text{card}\{M_3\}$.

Denote the $[24, 24]$ binary code by $\aleph$, and define

$$\lambda(\mathbf{u}, \mathbf{v}) = \begin{cases} 1 & \text{if } \|\mathbf{u}, \mathbf{v}\|_H < 4, \\ \frac{1}{6} & \text{if } \|\mathbf{u}, \mathbf{v}\|_H = 4, \\ 0 & \text{otherwise.} \end{cases}$$

Without loss of generality, if $\mathbf{u} = (0, 0, \ldots, 0)$, $\|\mathbf{u}, \mathbf{v}\|_H = \|\mathbf{u}, \mathbf{w}\|_H = 4$, and $\|\mathbf{v}, \mathbf{w}\|_H = 8$, then $v_i w_i = 0$ for all indices i. Thus, it is easy to see that

$$\sum_{\mathbf{u} \in \aleph_3} \lambda(\mathbf{u}, \mathbf{v}) \leq 1$$

for every codeword $\mathbf{v} \in \aleph$, and therefore

$$\sum_{\mathbf{u} \in \aleph_3} \sum_{\mathbf{v} \in \aleph} \lambda(\mathbf{u}, \mathbf{v}) = \sum_{\mathbf{v} \in \aleph} \sum_{\mathbf{u} \in \aleph_3} \lambda(\mathbf{u}, \mathbf{v}) \leq 2^{24}.$$

On the other hand,

$$\sum_{\mathbf{v} \in \aleph} \lambda(\mathbf{u}, \mathbf{v}) = 1 + \binom{24}{1} + \binom{24}{2} + \binom{24}{3} + \frac{1}{6}\binom{24}{4} = 2^{12}.$$

Then, we have

$$\text{card}\{\aleph_3\} \leq 2^{24}/2^{12} = 2^{12}.$$

Also, at most 24 points of M_3 correspond to one codeword of $\aleph_3$. Thus,

$$\text{card}\{M_3\} \leq 24 \times 2^{12} = 98304. \tag{9.21}$$

From our consideration of these cases, and since

$$97152 + 1104 + 98304 = 196560,$$

we conclude that equality must hold simultaneously in (9.19), (9.20), and (9.21). Then, by Case 3 and choosing an appropriate basis for E^{24}, M_3 contains all the points of $[-3^1|1^{23}]^*/\sqrt{8}$ type, where a point $\mathbf{x}$ is of $[a^j|b^k|\cdots]^*$ type if $x_i = a$ for j coordinates, $x_i = b$ for k coordinates, etc. In this case, $M(\Lambda^*)$ contains no point of $[-2^1|2^7|0^{16}]^*/\sqrt{8}$ type. Otherwise, one could find a point of $[1^{24}]/\sqrt{8}$ type in Λ^*, which is impossible. Thus, denoting by M^* the set of 759 points of $[2^8|0^{16}]^*/\sqrt{8}$ type corresponding to $\aleph_1$ and using (9.19), we have $M^* \subset M(\Lambda^*)$. Now we proceed to show that $M(\Lambda^*)$ can be generated by $\mathbf{u}_{24}^* = (-3, 1, \ldots, 1)/\sqrt{8}$, M^*, and G_{24}.

Let $\mathbf{u}^* = (u_1^*, u_2^*, \ldots, u_{24}^*)/\sqrt{8} \in \Lambda^*$ be a point of $[2^k|0^{24-k}]/\sqrt{8}$ type, $k \geq 8$. It follows from Case 1 that there is a point

$$\mathbf{v}^* = (v_1^*, v_2^*, \ldots, v_{24}^*)/\sqrt{8} \in M^*$$

such that

$$|u_i^*| = |v_i^*| = 2$$

for l indices with $5 \leq l \leq 8$. Then one can find a point $\mathbf{w}^* \in G_{24}$ such that $\mathbf{u}^* + \mathbf{v}^* + \mathbf{w}^*$ is of $[2^{k+8-2l}|0^{16+2l-k}]/\sqrt{8}$ type. Clearly, $k + 8 - 2l < k$. By reduction, $\mathbf{u}^*$ can be generated by M^* and G_{24}. From this it is clear that $M(\Lambda^*)$ is generated by $\mathbf{u}_{24}^*$, M^*, and G_{24}. Comparing this with the basis of Λ_{24} defined in Section 1, it follows that $M(\Lambda^*)$ is isometric to $M(\Lambda_{24})$. Theorem 9.4 is proved. $\square$

10. Multiple Sphere Packings

10.1. Introduction

A family $S_n + X$ of unit spheres is said to form a *k-fold packing* in E^n if each point of the space belongs to the interiors of at most k spheres of the family. In particular, when X is a lattice, the corresponding family is called a k-fold lattice packing. Let $m(S_n, k, l)$ be the maximal number of unit spheres forming a k-fold packing and contained in lI_n. Then analogously to the densities of classical sphere packings we define

$$\delta_k(S_n) = \limsup_{l \to +\infty} \frac{m(S_n, k, l)v(S_n)}{v(lI_n)}$$

and

$$\delta_k^*(S_n) = \sup_\Lambda \frac{v(S_n)}{\det(\Lambda)},$$

where the supremum is over all lattices Λ such that $S_n + \Lambda$ is a k-fold lattice packing in E^n.

Usually, $\delta_k(S_n)$ and $\delta_k^*(S_n)$ are called the density of the densest k-fold sphere packings and the density of the densest k-fold lattice sphere packings, respectively. Clearly, one can define $\delta_k(K)$ and $\delta_k^*(K)$ in a similar way for a general convex body K. Also, for these densities, it is easy to see that both

$$k\delta(K) \le \delta_k(K) \le k \tag{10.1}$$

and

$$k\delta^*(K) \le \delta_k^*(K) \le \delta_k(K) \tag{10.2}$$

hold for every convex body K and for every index k. In this chapter we will discuss some basic results about the densities $\delta_k(S_n)$ and $\delta_k^*(S_n)$.

Remark 10.1. *Obviously, k-fold packings are not always packings. Nevertheless, this generalization is very natural and has attracted the attention of many geometers.*

10.2. A Basic Theorem of Asymptotic Type

When n is fixed and $k \to \infty$, we have the following basic theorem describing the asymptotic behavior of $\delta_k(S_n)$ and $\delta_k^*(S_n)$.

Theorem 10.1.

$$\lim_{k \to \infty} \frac{\delta_k(S_n)}{k} = \lim_{k \to \infty} \frac{\delta_k^*(S_n)}{k} = 1.$$

Although this is rather counterintuitive when compared with known results about $\delta(S_n)$ and $\delta^*(S_n)$, this theorem itself is easy to prove. Nevertheless, for the convenience of further generalization, we will deduce it from some stronger and more general results.

Lemma 10.1. *Let K be a convex body, Λ a lattice, and let D be a polytope such that $D + \Lambda$ is a tiling of E^n. If there exist k points $\mathbf{p}_1, \mathbf{p}_2, \ldots, \mathbf{p}_k$ in Λ such that*

$$K \subset \bigcup_{i=1}^{k} (D + \mathbf{p}_i), \tag{10.3}$$

then

$$\delta_k^*(K) \geq \frac{v(K)}{v(D)}.$$

Proof. If a point $\mathbf{x} \in E^n$ belongs to the interiors of m translates of K in $K + \Lambda$, say $K + \mathbf{q}_1, K + \mathbf{q}_2, \ldots, K + \mathbf{q}_m$, then

$$\mathbf{x} - \mathbf{q}_j \in \mathrm{int}(K) \tag{10.4}$$

holds for $j = 1, 2, \ldots, m$. By (10.3) and (10.4) it follows that

$$\mathbf{x} - \mathbf{q}_1 \in D + \mathbf{p}_i$$

for some index i, $1 \leq i \leq k$. Hence, writing

$$D_j = D + \mathbf{p}_i + \mathbf{q}_1 - \mathbf{q}_j,$$

it follows that

$$\mathbf{x} - \mathbf{q}_j \in D_j$$

for $j = 1, 2, \ldots, m$. Therefore, by (10.4) we have

$$D_j \cap \text{int}(K) \neq \emptyset$$

for $j = 1, 2, \ldots, m$. On the other hand, it is clear that D_j is a tile in the tiling $D + \Lambda$. Thus, each D_j is some $D + \mathbf{p}_i$, and therefore

$$m \leq k.$$

It follows that $K + \Lambda$ is a k-fold lattice packing with density $v(K)/v(D)$. Lemma 10.1 is proved.   □

Lemma 10.2.
$$\delta_k^*(K) \geq k - c(K)k^{(n-1)/n},$$

where $c(K)$ is a positive constant depending only on K.

Proof. As usual, we denote the *Minkowski sum* of two convex bodies K_1 and K_2 by $K_1 + K_2$. In other words,

$$K_1 + K_2 = \{\mathbf{x} + \mathbf{y} : \mathbf{x} \in K_1, \mathbf{y} \in K_2\}.$$

We will require *Steiner's formula*, that

$$v(K_1 + tK_2) = v(K_1) + \sum_{i=1}^{n} \binom{n}{i} W_i(K_1, K_2) t^i \qquad (10.5)$$

holds for all positive numbers t, where $W_i(K_1, K_2)$ (the ith *mixed volume* of K_1 and K_2) are positive constants depending only on K_1, K_2, and i. In particular,

$$W_n(K_1, K_2) = v(K_2).$$

Taking $\Lambda = tZ_n$ and $D = tI_n$, it is obvious that $D + \Lambda$ is a tiling and the tiles that intersect K are contained in $K + 2D$. Thus, by (10.5), K is covered by

$$\left\lfloor \frac{v(K + 2D)}{v(D)} \right\rfloor = \left\lfloor \frac{v(K)}{t^n} + \sum_{i=1}^{n} 2^i \binom{n}{i} W_i(K, I_n) t^{i-n} \right\rfloor$$

tiles of the tiling.

Write

$$f(t) = \frac{v(K)}{t^n} + \sum_{i=1}^{n} 2^i \binom{n}{i} W_i(K, I_n) t^{i-n}.$$

Clearly, the function $f(t)$ is continuous on $(0, \infty)$.

$$\lim_{t \to 0} f(t) = \infty,$$

and

$$\lim_{t \to \infty} f(t) = 2^n.$$

Thus, for every large number k there exists a positive number t_k such that

$$k = \frac{v(K)}{t_k^n} + \sum_{i=1}^{n} 2^i \binom{n}{i} W_i(K, I_n) t_k^{i-n}. \qquad (10.6)$$

By Lemma 10.1, $K + t_k Z_n$ forms a k-fold lattice packing with density

$$\frac{v(K)}{t_k^n} = k - \sum_{i=1}^{n} 2^i \binom{n}{i} W_i(K, I_n) t_k^{i-n}. \qquad (10.7)$$

By (10.6), we have

$$k \geq \frac{v(K)}{t_k^n}$$

and therefore

$$t_k \geq \left(\frac{v(K)}{k} \right)^{1/n}. \qquad (10.8)$$

Then, it follows from Lemma 10.1, (10.7), and (10.8) that

$$\delta_k^*(K) \geq \frac{v(K)}{v(D)} = \frac{v(K)}{t_k^n} \geq k - c(K) k^{(n-1)/n},$$

where $c(K)$ is a suitable positive constant depending only upon K. Lemma 10.2 is proved.   □

Proof of Theorem 10.1. Theorem 10.1 is a direct consequence of (10.1), (10.2), and Lemma 10.2.   □

Remark 10.2. *Theorem 10.1 can be found in Bolle [1] and in Groemer [4]. In fact, this theorem is true not only for spheres but also for every convex body, and even for any bounded subset of E^n with positive Jordan measure.*

Remark 10.3. *In the two-dimensional case, using a more explicit method, Bolle [3] proved*

$$k - c_1 k^{2/5} \leq \delta_k^*(S_2) \leq k - c_2 k^{1/4},$$

where c_1 and c_2 are suitable positive constants.

Remark 10.4. *For general n and k, by modifying Blichfeldt's method, Few [3] and [6] proved*

$$\delta_k(S_n) \leq \left((n+1)^k - 1 \right) \left(1 + \frac{1}{n} \right) \left(\frac{k}{k+1} \right)^{n/2}$$

and, in particular,

$$\delta_2(S_n) \leq \frac{4(n+2)}{3} \left(\frac{2}{3}\right)^{n/2}.$$

10.3. A Theorem of Few and Kanagasahapathy

Theorem 10.2 (Few and Kanagasahapathy [1]).

$$\delta_2^*(S_3) = \frac{8\pi}{9\sqrt{3}}.$$

In addition, the optimal lattice is unique up to isometry.

Taking Λ^* to be the lattice with basis $\mathbf{a}_1 = (\sqrt{3}, 0, 0)$, $\mathbf{a}_2 = (\sqrt{3}/2, 3/2, 0)$, and $\mathbf{a}_3 = (0, 0, 1)$, routine computation shows that $S_3 + \Lambda^*$ is a double packing with density $8\pi/9\sqrt{3}$. Thus,

$$\delta_2^*(S_3) \geq \frac{8\pi}{9\sqrt{3}}. \tag{10.9}$$

Let Λ be any lattice in E^3 such that $S_3 + \Lambda$ forms a double packing. We proceed to show that

$$\det(\Lambda) \geq \frac{3\sqrt{3}}{2}, \tag{10.10}$$

where equality holds only if Λ is isometric to Λ^*. We will require the following two technical lemmas concerning Λ.

Lemma 10.3. *Let* $\mathbf{opq}$, *where* $\mathbf{p}$, $\mathbf{q} \in \Lambda$, *be an acute-angled triangle in which* $\|\mathbf{o}, \mathbf{p}\| \leq 2$ *is the shortest side. Assume that* $\mathbf{p} = (p, 0, 0)$ *and* $\mathbf{q} = (\alpha, q, 0)$, *where* α, p, *and* q *are all positive. Then, we have* $p \geq 1$,

$$\|\mathbf{o}, \mathbf{q}\|^2 + \|\mathbf{p}, \mathbf{q}\|^2 \geq 6, \tag{10.11}$$

$$q \geq \sqrt{4 - p^2}, \tag{10.12}$$

$$q \geq \frac{p^2}{2}\sqrt{4 - p^2}, \quad 2 \leq p^2 \leq 3, \tag{10.13}$$

and

$$q \geq \frac{\sqrt{3}p}{2}, \quad p^2 \geq 3, \tag{10.14}$$

where equality in (10.12) holds only if $\alpha = 0$ or $\alpha = p$.

Proof. First, we must have $p \geq 1$, since otherwise **o** would be covered three times by the interiors of $S_3 - \mathbf{p}$, S_3, and $S_3 + \mathbf{p}$.

Take $\mathbf{x} = (p/2, \sqrt{4 - p^2}/2, 0)$. Then $\mathbf{q} \notin \text{int}(S_3) + \mathbf{x}$, since otherwise some point near $\mathbf{x}$ would be covered three times. Hence, (10.12) follows from the fact that **opq** is an acute-angled triangle.

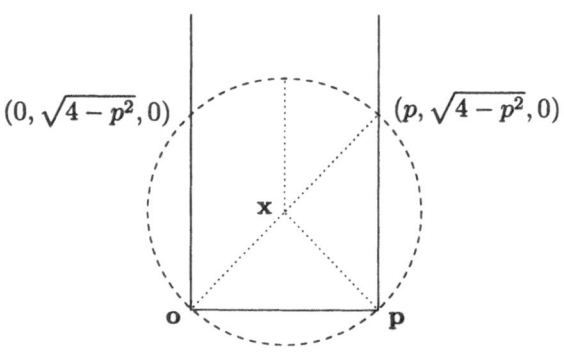

$$(0, \sqrt{4 - p^2}, 0) \qquad\qquad (p, \sqrt{4 - p^2}, 0)$$

Figure 10.1

If $p^2 \geq 3$, then (10.11) is trivial, and if $p^2 \leq 2$, then $\|\mathbf{o}, \mathbf{q}\|^2 + \|\mathbf{p}, \mathbf{q}\|^2$ attains its minimum when $\mathbf{q} = (0, \sqrt{4 - p^2}, 0)$ or $\mathbf{q} = (p, \sqrt{4 - p^2}, 0)$ (see Figure 10.1). In either case

$$\|\mathbf{o}, \mathbf{q}\|^2 + \|\mathbf{p}, \mathbf{q}\|^2 = 4 + (4 - p^2) \geq 6.$$

Suppose now that $2 \leq p^2 \leq 3$. In this case both $\|\mathbf{o}, \mathbf{q}\|^2 + \|\mathbf{p}, \mathbf{q}\|^2$ and q attain their minima when $\|\mathbf{o}, \mathbf{q}\| = p$ and $\|\mathbf{q}, \mathbf{x}\| = 1$ (see Figure 10.1). Writing $\theta = \angle \mathbf{poq}$, we have

$$\sin \frac{\theta}{2} = \frac{\sqrt{4 - p^2}}{2}, \qquad \cos \frac{\theta}{2} = \frac{p}{2},$$

and

$$q \geq p \sin \theta = 2p \sin \frac{\theta}{2} \cos \frac{\theta}{2} = \frac{p^2}{2} \sqrt{4 - p^2},$$

which proves (10.13). Also,

$$\|\mathbf{p}, \mathbf{q}\| = 2p \sin \frac{\theta}{2} = p\sqrt{4 - p^2}$$

and

$$\|\mathbf{o}, \mathbf{q}\|^2 + \|\mathbf{p}, \mathbf{q}\|^2 - 6 = p^2 + p^2(4 - p^2) - 6$$
$$= (2 - p^2)(p^2 - 3) \geq 0,$$

which completes the proof of (10.11).

The result (10.14) is trivial, since $\|o, q\| \geq p$ and $\|p, q\| \geq p$. Lemma 10.3 is proved. □

Lemma 10.4. *Let* **y** *be any point in the parallelogram with vertices* **o**, **p**, $\frac{1}{2}(\mathbf{p} + \mathbf{q})$, *and* $\frac{1}{2}(\mathbf{q} - \mathbf{p})$. *Then either*

$$\|\mathbf{y}, \tfrac{1}{2}\mathbf{p}\| \leq \tfrac{1}{2}\|\mathbf{o}, \mathbf{q}\| \quad or \quad \|\mathbf{y}, \tfrac{1}{2}\mathbf{q}\| \leq \tfrac{1}{2}\|\mathbf{o}, \mathbf{p}\|.$$

Proof. Let **v** be the perpendicular foot from $\frac{1}{2}(\mathbf{p}+\mathbf{q})$ to the line determined by **o** and $\mathbf{q} - \mathbf{p}$. Then, since $\|\mathbf{o}, \mathbf{q}\| \geq \|\mathbf{o}, \mathbf{p}\|$, **v** is between **o** and $\frac{1}{2}(\mathbf{q} - \mathbf{p})$ (see Figure 10.2). Since $\angle(\mathbf{q} - \mathbf{p})\mathbf{v}(\mathbf{p} + \mathbf{q}) = \pi/2$ and $\frac{1}{2}\mathbf{q}$ is the midpoint of $\frac{1}{2}(\mathbf{q} - \mathbf{p})$ and $\frac{1}{2}(\mathbf{p} + \mathbf{q})$, we have

$$\|\mathbf{v}, \tfrac{1}{2}\mathbf{q}\| = \tfrac{1}{2}\|\mathbf{o}, \mathbf{p}\|.$$

Similarly, since $\frac{1}{2}\mathbf{p}$ is the midpoint of $\frac{1}{2}(\mathbf{p} + \mathbf{q})$ and $\frac{1}{2}(\mathbf{p} - \mathbf{q})$,

$$\|\mathbf{v}, \tfrac{1}{2}\mathbf{p}\| = \tfrac{1}{2}\|\mathbf{o}, \mathbf{q}\|.$$

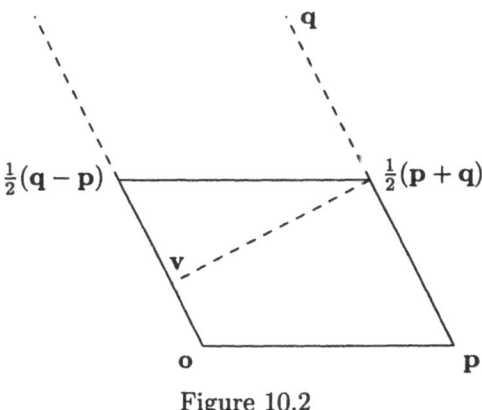

Figure 10.2

Hence, if **y** belongs to the triangle with vertices $\frac{1}{2}(\mathbf{p} + \mathbf{q})$, $\frac{1}{2}(\mathbf{q} - \mathbf{p})$, and **v**, we have

$$\|\mathbf{y}, \tfrac{1}{2}\mathbf{q}\| \leq \tfrac{1}{2}\|\mathbf{o}, \mathbf{p}\|,$$

and if **y** belongs to the quadrilateral with vertices **v**, **o**, **p**, and $\frac{1}{2}(\mathbf{p} + \mathbf{q})$, we have

$$\|\mathbf{y}, \tfrac{1}{2}\mathbf{p}\| \leq \tfrac{1}{2}\|\mathbf{o}, \mathbf{q}\|.$$

Lemma 10.4 is proved. □

Proof of Theorem 10.2. First we try to prove (10.10). Let **p** be one of the nearest points of Λ to **o**. Then $\|\mathbf{o}, \mathbf{p}\| \geq 1$, since otherwise **o** would be covered three times by the interiors of $S_3 - \mathbf{p}$, S_3, and $S_3 + \mathbf{p}$. On the other hand, we may suppose that $\|\mathbf{o}, \mathbf{p}\| < 2$; otherwise, $S_3 + \Lambda$ is a normal packing.

Let L denote the line determined by $\mathbf{o}$ and $\mathbf{p}$. Let $\mathbf{q}_1 \in \Lambda \setminus L$ be a point such that the distance from $\mathbf{q}_1$ to L is minimal, and let $\mathbf{q}_2 \in \Lambda \setminus L$ be a point such that $\|\mathbf{o}, \mathbf{q}_2\| = r$ is minimal. We may take $\mathbf{p} = (p, 0, 0)$ with $1 \leq p < 2$, and $\mathbf{q}_i = (\alpha_i, q_i, 0)$ with $q_i \geq 0$ and $0 \leq \alpha_i < p/2$. Since $\mathbf{op}$ is the shortest side of the triangle $\mathbf{opq}_i$, and $0 \leq \alpha_i < p/2$, it follows that the triangles $\mathbf{opq}_1$ and $\mathbf{opq}_2$ are acute-angled. (We are not assuming that $\mathbf{o}$, $\mathbf{p}$, $\mathbf{q}_1$, and $\mathbf{q}_2$ are coplanar. We make two choices of axes of coordinates, one for each basis.)

Let $\mathbf{u}_i$ ($i = 1, 2$) be chosen such that $\mathbf{p}$, $\mathbf{q}_i$, and $\mathbf{u}_i$ generate Λ. We may take $\mathbf{u}_i = (a_i, b_i, c_i)$, where the corresponding points $\mathbf{u}_i^* = (a_i, b_i, 0)$ belong to the parallelogram with vertices $\mathbf{o}$, $\mathbf{p}$, $\frac{1}{2}(\mathbf{p} + \mathbf{q}_i)$, and $\frac{1}{2}(\mathbf{q}_i - \mathbf{p})$ (see Figure 10.3).

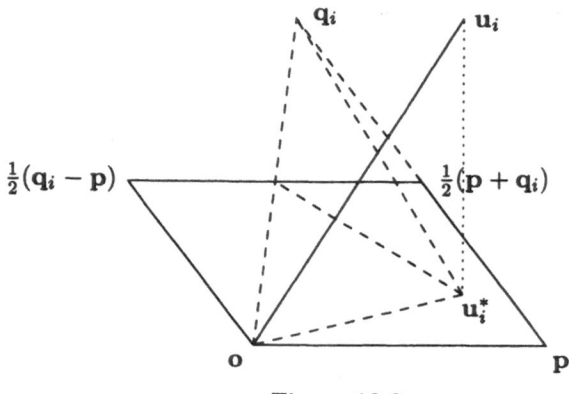

Figure 10.3

Based on the choices of $\mathbf{q}_1$ and $\mathbf{u}_1$ it follows that

$$b_1^2 + c_1^2 \geq q_1^2, \quad 0 \leq b_1 \leq \frac{q_1}{2},$$

and therefore

$$c_1^2 \geq q_1^2 - b_1^2 \geq \frac{3q_1^2}{4}.$$

Thus, we have

$$\det(\Lambda) = pq_1 c_1 \geq \frac{\sqrt{3}pq_1^2}{2}. \tag{10.15}$$

We now consider five cases.

Case 1. $p \leq 1.3$. By (10.15), (10.12), and routine computation it follows that

$$\det(\Lambda) \geq \frac{\sqrt{3}p(4 - p^2)}{2} \geq \frac{3\sqrt{3}}{2},$$

where equalities hold if and only if the considered lattice Λ is isometric to Λ^*.

Case 2. $2.1 \leq p^2 \leq 3$. By (10.15), (10.13), and routine computation we have

$$\det(\Lambda) \geq \frac{\sqrt{3}p^5(4-p^2)}{8} > \frac{3\sqrt{3}}{2}.$$

Case 3. $3 \leq p^2 \leq 4$. By (10.15) and (10.14),

$$\det(\Lambda) \geq \frac{3\sqrt{3}}{8}p^3 \geq \frac{27}{8} > \frac{3\sqrt{3}}{2}.$$

Case 4. $1.3 \leq p \leq \sqrt{2.1}$ and $r^2 \geq 3$. Let $\mathbf{w}_2$ be the lattice point in the plane $\{\mathbf{x} : x_3 = c_2\}$ that is nearest to $\mathbf{o}$. Then,

$$\|\mathbf{o}, \mathbf{w}_2\|^2 \leq c_2^2 + \frac{p^2}{4} + \frac{q_2^2}{4} \leq c_2^2 + \frac{p^2}{4} + \frac{r^2}{4}.$$

Hence, based on the choice of $\mathbf{q}_2$,

$$c_2^2 + \frac{p^2}{4} + \frac{r^2}{4} \geq r^2,$$

$$c_2^2 \geq \frac{3r^2 - p^2}{4} \geq \frac{9 - p^2}{4}.$$

Therefore, by (10.12) and routine analysis we have

$$\det(\Lambda) = pq_2c_2 \geq \frac{p}{2}\sqrt{(4-p^2)(9-p^2)} > \frac{3\sqrt{3}}{2}.$$

Case 5. $1.3 \leq p \leq \sqrt{2.1}$ and $r^2 \leq 3$. The triangle $\mathbf{oq}_2\mathbf{u}_2$ is acute-angled and $\mathbf{oq}_2$ is the shortest side. Hence, by elementary geometry and (10.11),

$$\|\mathbf{o}, \mathbf{u}_2\|^2 + \|\mathbf{u}_2, \mathbf{q}_2\|^2 = 2c_2^2 + \|\mathbf{o}, \mathbf{u}_2^*\|^2 + \|\mathbf{q}_2, \mathbf{u}_2^*\|^2$$
$$= 2c_2^2 + 2\|\mathbf{u}_2^*, \tfrac{1}{2}\mathbf{q}_2\|^2 + \tfrac{1}{2}r^2 \qquad (10.16)$$
$$\geq 6.$$

We now need to prove that

$$c_2^2 \geq \frac{9 - p^2}{4}. \qquad (10.17)$$

If $\|\mathbf{u}_2^*, \tfrac{1}{2}\mathbf{q}_2\| \leq p/2$, then (10.17) follows from (10.16). If $\|\mathbf{u}_2^*, \tfrac{1}{2}\mathbf{q}_2\| > p/2$, it follows from Lemma 10.4 that

$$\|\mathbf{u}_2^*, \tfrac{1}{2}\mathbf{p}\| \leq \tfrac{1}{2}r. \qquad (10.18)$$

If the angle $\angle u_2 op$ is obtuse, then the angle $\angle sop$, where $s = u_2 + p$, is acute and

$$\|s^*, \tfrac{1}{2}p\| \leq \|u_2^*, \tfrac{1}{2}p\| \leq \tfrac{1}{2}r, \tag{10.19}$$

where $s^* = u_2^* + p$. Hence applying (10.11) to the triangle opt, where $t = u_2$ or s,

$$\begin{aligned}
\|o, t\|^2 + \|t, p\|^2 &= 2c_2^2 + \|o, t^*\|^* + \|p, t^*\|^2 \\
&= 2c_2^2 + 2\|t^*, \tfrac{1}{2}p\|^2 + \tfrac{1}{2}p^2 \\
&\geq 6,
\end{aligned} \tag{10.20}$$

where $t^* = u_2^*$ or s^*. In this case, (10.17) follows from (10.18), (10.19), and (10.20). Then, by (10.12) and (10.17),

$$\det(\Lambda) = pq_2 c_2 \geq \frac{p}{2}\sqrt{(4 - p^2)(9 - p^2)} > \frac{3\sqrt{3}}{2}.$$

Hence, (10.10) is proved. Clearly, (10.9) and (10.10) together yield Theorem 10.2. $\qquad\square$

Remark 10.5. *Theorem 10.2 is the only known result about the exact values of $\delta_k(S_n)$ and $\delta_k^*(S_n)$ when $n \geq 3$ and $k \geq 2$.*

10.4. Remarks on Multiple Circle Packings

In E^2, much more explicit results are available. Since their proofs are elementary and technical, we only quote the results.

k	$\delta_k^*(S_2)$	Author
2	$\frac{\pi}{\sqrt{3}}$	Heppes [1]
3	$\frac{3\pi}{2}$	Heppes [1]
4	$\frac{2\pi}{\sqrt{3}}$	Heppes [1]
5	$\frac{2\pi}{\sqrt{7}}$	Blundon [2]
6	$\frac{35\pi}{8\sqrt{6}}$	Blundon [2]
7	$\frac{8\pi}{\sqrt{15}}$	Bolle [2]
8	$\frac{3969\pi}{4\sqrt{(220 - 2\sqrt{193})(449 + 32\sqrt{193})}}$	Yakovlev [1]

Several Vienna dissertations dealt with problems of this kind. Unfortunately, it is not easy to obtain them. Also, Linhart [1] and Temesvári, Horváth, and Yakovlev [1] developed algorithms with which one can determine $\delta_k^*(S_2)$ for each k to any prescribed accuracy.

Multiple packings of general 2-dimensional convex domains have been studied by many authors. For results of this kind we refer to G. Fejes Tóth and W. Kuperberg [1].

11. Holes in Sphere Packings

11.1. Spherical Holes in Sphere Packings

In order to study the efficiency of a sphere packing, it is both important and interesting to investigate its holes, especially spherical holes. Let $S_n + X$ be a sphere packing in E^n, and let $r(S_n, X)$ be the supremum of the radii of all spheres disjoint from any sphere of the packing. The number $1/r(S_n, X)$ is called the *closeness* of the packing $S_n + X$. We define

$$r(S_n) = \inf_X \{r(S_n, X)\},$$

where the infimum is taken over all sets X such that $S_n + X$ is a packing. By routine argument it follows that there exists a discrete set X such that

$$r(S_n) = r(S_n, X).$$

The corresponding packing $S_n + X$ is called the *closest sphere packing*. Similarly, one can define $r^*(S_n)$ by assuming that X is a lattice. Clearly, we have

$$r(S_n) \le r^*(S_n).$$

These concepts were first introduced by L. Fejes Tóth [11], who suggested the problems of determining the number $r(S_n)$ and the corresponding closest sphere packings. These problems, like Kepler's conjecture and the Gregory-Newton problem, are both natural and important. However, our knowledge about them is comparatively limited. In this section we will prove the following main result.

Theorem 11.1 (Böröczky [2]).

$$r(S_3) = r^*(S_3) = \sqrt{5/3} - 1.$$

Also, up to isometry,

$$r(S_3, X) = r(S_3) = r^*(S_3)$$

if and only if X is a suitable space-centered cubic lattice.

To prove this theorem, we need several technical lemmas.

Lemma 11.1 (Böröczky [2]). *Let $T = \mathbf{v}_1\mathbf{v}_2\mathbf{v}_3\mathbf{v}_4$ be a tetrahedron with circumradius r such that $\|\mathbf{v}_i, \mathbf{v}_j\| \geq 2$, $i \neq j$, and the dihedral angle at edge $\mathbf{v}_1\mathbf{v}_2$ is at most $\pi/3$. Then*

$$r \geq \sqrt{5/3},$$

where equality holds if and only if T is congruent to the tetrahedron $T^* = \mathbf{v}_1^*\mathbf{v}_2^*\mathbf{v}_3^*\mathbf{v}_4^*$ for which

$$\|\mathbf{v}_1^*, \mathbf{v}_3^*\| = \|\mathbf{v}_2^*, \mathbf{v}_4^*\| = 4/\sqrt{3}$$

and

$$\|\mathbf{v}_1^*, \mathbf{v}_2^*\| = \|\mathbf{v}_2^*, \mathbf{v}_3^*\| = \|\mathbf{v}_3^*, \mathbf{v}_4^*\| = \|\mathbf{v}_4^*, \mathbf{v}_1^*\| = 2.$$

Proof. Let S^* be the circumsphere of T^* centered at $\mathbf{o}$. Lemma 11.1 will be proved by showing that any tetrahedron T inscribed in S^* and satisfying the conditions of the lemma is congruent to T^*.

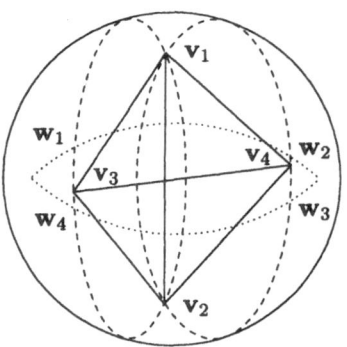

Figure 11.1

Assume that $\mathbf{v}_1\mathbf{v}_2$ is parallel to $\mathbf{v}_1^*\mathbf{v}_2^*$, and the four points $\mathbf{v}_1$, $\mathbf{v}_1^*$, $\mathbf{v}_2^*$, and $\mathbf{v}_2$ are contained in a half great circle of S^*. Denote by G the part of $\mathrm{bd}(S^*)$ cut out by the dihedral angle of T at the edge $\mathbf{v}_1\mathbf{v}_2$, by H the spherical set $\{\mathbf{x} \in \mathrm{bd}(S^*) : \|\mathbf{x}, \mathbf{v}_1\| \geq 2, \|\mathbf{x}, \mathbf{v}_2\| \geq 2\}$, and by Q the intersection of G and H. Clearly, we have

$$\{\mathbf{v}_3, \mathbf{v}_4\} \subset Q, \tag{11.1}$$

and Q is a spherical domain with four vertices, say $\mathbf{w}_1$, $\mathbf{w}_2$, $\mathbf{w}_3$, and $\mathbf{w}_4$ in a circular order. Also,

$$d(Q) = \|\mathbf{w}_1, \mathbf{w}_3\| = \|\mathbf{w}_2, \mathbf{w}_4\|, \tag{11.2}$$

where $\{\mathbf{w}_1, \mathbf{w}_3\}$ and $\{\mathbf{w}_2, \mathbf{w}_4\}$ are the only solutions (see Figure 11.1).

Since (11.1) and (11.2) hold, we may modify the considered tetrahedron as follows.

Step 1. Replace $\mathbf{v}_1$ and $\mathbf{v}_2$ by $\mathbf{v}_1^*$ and $\mathbf{v}_2^*$, respectively. Then, replace $\mathbf{v}_3$ and $\mathbf{v}_4$ by the new points $\mathbf{w}_1$ and $\mathbf{w}_3$, respectively.

Step 2. Move $\mathbf{v}_3$ and $\mathbf{v}_4$ along the boundary of H until the dihedral angle of T at $\mathbf{v}_1\mathbf{v}_2$ is $\pi/3$.

Clearly, $d(Q)$ strictly increases during this process.

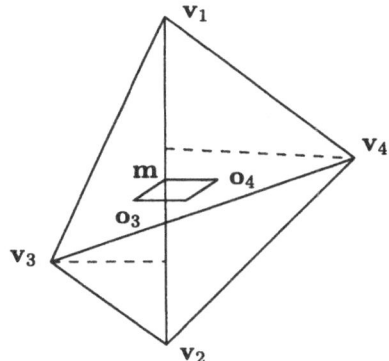

Figure 11.2

Let T be a modified tetrahedron, $\mathbf{m}$ the midpoint of $\mathbf{v}_1\mathbf{v}_2$, $\mathbf{o}_i$ the center of the circumcircle of $\mathbf{v}_1\mathbf{v}_2\mathbf{v}_i$, and write $x = \|\mathbf{m}, \mathbf{o}_3\|$ and $y = \|\mathbf{m}, \mathbf{o}_4\|$. Then, elementary but complicated computation (see Figure 11.2) yields that

$$\frac{\sqrt{3}+1}{6} \le xy \le \frac{1}{2}$$

and

$$\|\mathbf{v}_3, \mathbf{v}_4\|^2 = 36 - \frac{96(1 + xy)}{2x^2y^2 + 2xy + 3} \le 4,$$

where equality holds if and only if T is congruent to T^*. Hence, Lemma 11.1 is proved. $\qquad\square$

Write $b = 4/\sqrt{3}$ and let Λ be the space-centered cubic lattice with basis $\mathbf{u}_1 = (b, 0, 0)$, $\mathbf{u}_2 = (0, b, 0)$, and $\mathbf{u}_3 = (b/2, b/2, b/2)$. Then, T^* is congruent to the tetrahedron with vertices $\mathbf{u}_1$, $\mathbf{u}_1 + \mathbf{u}_2$, $\mathbf{u}_3$, and $\mathbf{u}_3 + \mathbf{u}_1$. Also, since the Dirichlet-Voronoi cells of Λ are regular truncated octahedra, the solid angle formed by the three planes perpendicular to the edges of T^* meeting at a vertex is congruent to the solid angle at a vertex of a regular truncated octahedron.

Lemma 11.2 (Böröczky [2]). *There is no convex polyhedron bounded by one quadrangle and ten pentagons such that in each vertex of the polyhedron exactly three facets meet.*

Proof. Suppose there is such a polyhedron, say P. We proceed to deduce a contradiction. Let v_1, v_2, v_3, and v_4 be the vertices of the quadrangular facet of P, in a circular order (see Figure 11.3). Let w_i be the endpoint of the third edge of P emanating from v_i, $i = 1, \ldots, 4$. The points w_i are different from one another, and different from the points v_i, $i = 1, \ldots, 4$. For convenience, we write $v_5 = v_1$ and $w_5 = w_1$. Then, for $1 \leq i \leq 4$, the points w_i, v_i, v_{i+1}, and w_{i+1} are four consecutive vertices of a pentagonal facet of P. Let x_i be the fifth vertex of this facet. Again, the points v_i, w_i, x_i, $i = 1, \ldots, 4$, are all distinct. In the part of P considered so far the points v_i and w_i have valency 3, while x_i has valency 2. Thus, there is a third vertex, say y_i, that is connected by an edge to x_i. Again for convenience, we write $x_5 = x_1$ and $y_5 = y_1$. It is easy to see that $y_i \neq x_{i+1}$ and $y_i \neq y_{i+1}$, since otherwise we would obtain a triangular or quadrangular facet. If we have $y_i = x_{i+2}$ for $i = 1$ or 2, then y_{i+1} has valency 1; if we have $y_i = y_{i+2}$ for $i = 1$ or 2, then $y_i = y_{i+1} = y_{i+2}$. In the second case, $x_i w_{i+1} x_{i+1} y_i$ and $y_i x_{i+1} w_{i+2} x_{i+2}$ are quadrangular facets of P. Hence, the points y_i are different from one another and from the points v_i, w_i, and x_i, $i = 1, \ldots, 4$.

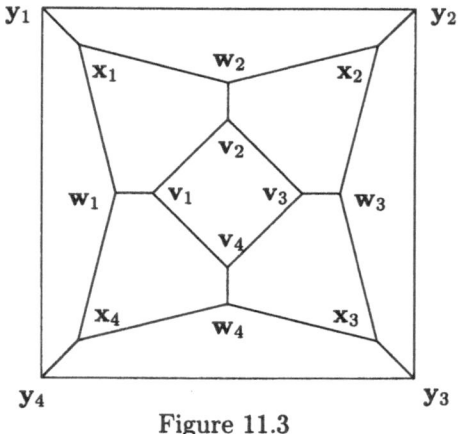

Figure 11.3

Now, y_i, x_i, w_{i+1}, x_{i+1}, and y_{i+1} are five consecutive vertices of a facet of P; thus the points y_i and y_{i+1} are connected by an edge. This means that all the vertices considered so far are trivalent, and our polyhedron has only ten facets, two quadrangles, and eight pentagons. By this contradiction Lemma 11.2 is proved.   □

It is well-known (see Coxeter [5]) that there are eight different types of regular polytopes in E^4, one of which, say $P^\star$, is bounded by 120 regular dodecahedra such that at each vertex four of them meet.

A *cell complex* $\{\mathcal{F}, \prec\}$ is a family $\mathcal{F}$ of cells (polytopes and their faces) with an incidence relation $\prec$ (for example, the relation $\subset$). Two cell complexes $\{\mathcal{F}_1, \prec_1\}$ and $\{\mathcal{F}_2, \prec_2\}$ are *isomorphic* if there is a one-to-one mapping

$$f: \quad \mathcal{F}_1 \mapsto \mathcal{F}_2$$

such that

$$f(X) \prec_2 f(Y)$$

if and only if $X \prec_1 Y$ for $X, Y \in \mathcal{F}_1$.

Lemma 11.3 (Böröczky [2]). *There is no tetravalent facet-to-facet tiling in E^3 such that each tile has exactly 12 pentagonal facets.*

Proof. Let $\wp$ be the family of all faces of $P^\star$. Then, $\{\wp, \subset\}$ is a cell complex. Without loss of generality, we label the facets of $P^\star$ as F_1, F_2, ..., F_{120} such that for $k = 1, 2, \ldots, 119$, the set

$$U_k = \bigcup_{i=1}^{k} F_i$$

is *homeomorphic* to S_3 and the set

$$U_k^\star = F_k \cap \left(\bigcup_{i=1}^{k-1} F_i \right)$$

is homeomorphic to S_2. For convenience, we write

$$\wp_k = \{X \in \wp : \ X \subset U_k\},$$

$$\wp_k^\star = \{X \in \wp : \ X \subset U_{k+1}^\star\},$$

and $\wp_{120} = \wp$.

Suppose, on the contrary, there is a tetravalent facet-to-facet tiling in E^3 such that each tile has exactly 12 pentagonal facets. Let P_1 be an arbitrary tile of the tiling, let $\Im_1$ be the family consisting of P_1 and all its faces, and let $\prec_1$ be the relation $\subset$. Clearly, the two cell complexes $\{\wp_1, \subset\}$ and $\{\Im_1, \prec_1\}$ are isomorphic. Let f_1 be an isomorphism between them. Now, for $k = 1, 2, \ldots, 120$, we proceed to define cell complexes $\{\Im_k, \prec_k\}$ and isomorphisms f_k between $\{\wp_k, \subset\}$ and $\{\Im_k, \prec_k\}$ by induction. In general, the cell complex $\{\Im_k, \prec_k\}$ has the following properties:

1. $\Im_k$ *is a family consisting of some tiles of the tiling and all their faces.*
2. *If $X \prec_k Y$, then $X \subset Y$.*

Suppose that we have a cell complex $\{\Im_k, \prec_k\}$ with the above properties and an isomorphism f_k between $\{\wp_k, \subset\}$ and $\{\Im_k, \prec_k\}$. Let $\{\Im_k^\star, \prec_k\}$ be the subcomplex of $\{\mathfrak{S}_k, \prec_k\}$ corresponding to $\{\wp_k^\star, \subset\}$ at the isomorphism f_k, let $Q_k \in \mathfrak{S}_k^\star$ be a facet of some tile, say $P_k \in \mathfrak{S}_k$, and let P_{k+1} be the tile such that $Q_k = P_k \cap P_{k+1}$. Since $U_{k+1}^\star$ is homeomorphic to S_2, all the facets in $\mathfrak{S}_k^\star$ are connected by edges of the tiles. On the other hand, since the tiling is tetravalent and is facet-to-facet, at every edge three tiles join properly together. Thus, P_{k+1} is independent of the choice of Q_k.

Let $\mathfrak{S}_{k+1}$ be the family consisting of $\mathfrak{S}_k$, P_{k+1}, and all the faces of P_{k+1}, and define the relation $\prec_{k+1}$ by

$$X \prec_{k+1} Y = \begin{cases} X \prec_k Y & \text{if } X, Y \in \mathfrak{S}_k, \\ X \subset Y & \text{if } Y \in \mathfrak{S}_{k+1} \setminus \mathfrak{S}_k. \end{cases}$$

Then, it is easy to show that the cell complex $\{\mathfrak{S}_{k+1}, \prec_{k+1}\}$ satisfies the required properties and is isomorphic to $\{\wp_{k+1}, \subset\}$.

As a result of this process, we obtain a cell complex $\{\mathfrak{S}_{120}, \prec_{120}\}$ that satisfies properties 1 and 2 and is isomorphic to $\{\wp, \subset\}$. Obviously, this contradicts the fact that $P_1, P_2, \ldots, P_{120}$ are tiles of the tiling. Thus, Lemma 11.3 is proved. $\square$

Now we introduce a fundamental result concerning cap packings and cap coverings in $\mathrm{bd}(S_3)$ that will be useful in the proof of Theorem 11.1.

Lemma 11.4 (L. Fejes Tóth [9]). *Let $\gamma(m)$ be the maximum spherical radius such that m caps of such radius can be packed into $\mathrm{bd}(S_3)$. Let $\gamma'(m)$ be the minimum spherical radius such that m caps of such radius can cover $\mathrm{bd}(S_3)$, and write*

$$\theta_m = \frac{m}{6(m-2)}\pi.$$

Then,

$$\gamma(m) \leq \arccos\left(\tfrac{1}{2}\operatorname{cosec}\theta_m\right)$$

and

$$\gamma'(m) \geq \arccos\left(\sqrt{\tfrac{1}{3}}\cot\theta_m\right),$$

where equalities hold only if $m = 3$, 4, 6, or 12.

Proof. Let $\Omega_1, \Omega_2, \ldots, \Omega_m$ be m caps, of spherical radius γ, centered at $\mathrm{o}_1, \mathrm{o}_2, \ldots, \mathrm{o}_m$, respectively, and let $D_1, D_2, \ldots, D_m$ be the corresponding spherical Dirichlet-Voronoi cells. It is easy to see that $\Omega_i \subset D_i$ if the caps form a packing in $\mathrm{bd}(S_3)$, $D_i \subset \Omega_i$ if they form a covering in $\mathrm{bd}(S_3)$, and that the Dirichlet-Voronoi cells form a tiling in $\mathrm{bd}(S_3)$. For convenience, we assume that the Ω_i are inscribed in the D_i in the packing case and that the D_i are inscribed in the Ω_i in the covering case. Let k be the number of edges of the tiling. By Euler's formula and the fact that at least three edges meet at every vertex, it can be deduced that

$$k \leq 3(m-2). \tag{11.3}$$

Joining each $\mathbf{o}_i$ to the vertices of D_i and its perpendicular feet on the edges of D_i by arcs, $\text{bd}(S_3)$ is divided into $4k$ spherical triangles. Consider one of them, say T, and assume that its spherical angle at the corresponding $\mathbf{o}_i$ is α. It is well-known that

$$s(T) = \alpha - \arcsin(\cos\gamma \cdot \sin\alpha) \qquad (11.4)$$

in the packing case, and

$$s(T) = \alpha - \arctan(\cos\gamma \cdot \tan\alpha) \qquad (11.5)$$

in the covering case. Since (11.4) is a convex function of α and (11.5) is a concave function of α when $0 < \alpha < \pi/2$, by Jensen's inequalities we have

$$4\pi = \sum T \geq 4k\left(\frac{m\pi}{2k} - \arcsin\left(\cos\gamma \cdot \sin\frac{m\pi}{2k}\right)\right)$$

and therefore

$$\cos\gamma \geq \frac{\sin\frac{(m-2)\pi}{2k}}{\sin\frac{m\pi}{2k}}$$

in the packing case. Also,

$$4\pi = \sum T \leq 4k\left(\frac{m\pi}{2k} - \arctan\left(\cos\gamma \cdot \tan\frac{m\pi}{2k}\right)\right),$$

and hence

$$\cos\gamma \leq \frac{\tan\frac{(m-2)\pi}{2k}}{\tan\frac{m\pi}{2k}}$$

in the covering case. Thus, by (11.3) we have

$$\cos\gamma(m) \geq \frac{\sin(\pi/6)}{\sin\theta_m} = \frac{1}{2}\operatorname{cosec}\theta_m \qquad (11.6)$$

in the packing case, and

$$\cos\gamma'(m) \leq \frac{\tan(\pi/6)}{\tan\theta_m} = \frac{1}{\sqrt{3}}\cot\theta_m \qquad (11.7)$$

in the covering case. Checking the cases of equality in every step, it is easy to see that equalities hold in (11.6) and (11.7) only if $m = 3$, 4, 6, or 12. Thus, Lemma 11.4 follows from (11.6) and (11.7). $\qquad\square$

With these preparations, Theorem 11.1 can be proved as follows.

Proof of Theorem 11.1. For convenience, we write $\tau = \sqrt{5/3}$. In E^3, let $X = \{\mathbf{x}_1, \mathbf{x}_2, \ \ldots \}$ be a discrete set such that $S_3 + X$ is a packing and $\tau S_3 + X$ is a covering, and let D_i be the Dirichlet-Voronoi cell associated with $\mathbf{x}_i$. Obviously, we have

$$S_3 + \mathbf{x}_i \subset D_i \subset \tau S_3 + \mathbf{x}_i, \quad i = 1, 2, \ldots. \qquad (11.8)$$

First, we proceed to show that the tiling formed by the Dirichlet-Voronoi cells is tetravalent. In other words, at every vertex four Dirichlet-Voronoi cells meet. Suppose, on the contrary, a vertex $\mathbf{v}$ belongs to five cells, say $D_1, D_2, \ldots, D_5$. Then

$$\mathbf{x}_i \in \mathrm{bd}(dS_3) + \mathbf{v}, \quad i = 1, 2, \ldots, 5,$$

where d is the distance between $\mathbf{v}$ and $\mathbf{x}_1$. It follows from routine combinatorial argument that two of these five points, say $\mathbf{x}_1$ and $\mathbf{x}_2$, satisfy

$$\angle \mathbf{x}_1 \mathbf{v} \mathbf{x}_2 \leq \pi/2.$$

Thus, since $\|\mathbf{x}_1, \mathbf{x}_2\| \geq 2$, we obtain

$$d = \|\mathbf{v}, \mathbf{x}_1\| \geq \operatorname{cosec}(\pi/4) = \sqrt{2} > \sqrt{5/3} = \tau,$$

which contradicts (11.8). As a simple consequence, at every vertex of every Dirichlet-Voronoi cell three edges meet.

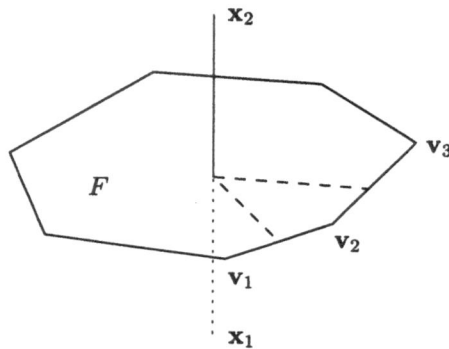

Figure 11.4

Second, we claim that none of the Dirichlet-Voronoi cells has a facet with more than six vertices. Suppose, on the contrary, that one of the cells, say D_1, has a facet F with more than six vertices $\mathbf{v}_1, \mathbf{v}_2, \ldots, \mathbf{v}_l$, and assume that $\mathbf{v}_2$ is also a vertex of $D_2, D_3,$ and D_4,

$$\angle \mathbf{v}_1 \mathbf{v}_2 \mathbf{v}_3 > 2\pi/3,$$

and $\mathbf{x}_1 \mathbf{x}_2$ is perpendicular to F. Then the dihedral angle of the tetrahedron $\mathbf{x}_1 \mathbf{x}_2 \mathbf{x}_3 \mathbf{x}_4$ at the edge $\mathbf{x}_1 \mathbf{x}_2$ is (see Figure 11.4)

$$\pi - \angle \mathbf{v}_1 \mathbf{v}_2 \mathbf{v}_3 < \pi/3.$$

Thus, Lemma 11.1 implies

$$\|\mathbf{x}_1, \mathbf{v}_2\| > \tau,$$

which contradicts (11.8).

Next, we are going to show that one of the Dirichlet-Voronoi cells, say D_1, has a hexagonal facet. Assume that D_1 has k facets with $e_1, e_2, \ldots, e_k$ edges, respectively. It follows from Euler's formula and the fact that three edges meet at every vertex that

$$\sum_{i=1}^{k} e_i = 6k - 12.$$

Thus, keeping the second claim in mind, the assertion can be easily shown if $k > 12$. We proceed to deduce the assertion by excluding the following cases.

Case 1. $k \leq 10$. It follows from Lemma 11.4 that there is a vertex $\mathbf{v}$ of D_1 such that

$$\|\mathbf{v}, \mathbf{x}_1\| \geq \sec \gamma'(10) > \tau,$$

which contradicts (11.8).

Case 2. $k = 11$ *and* $e_i \leq 5$ *for* $i = 1, 2, \ldots, 11$. By Euler's formula we have, without loss of generality, $e_1 = 4$ and $e_2 = \cdots = e_{11} = 5$. Then, by Lemma 11.2 this is impossible.

Case 3. *Each of the Dirichlet-Voronoi cells has exactly* 12 *facets, each of which has at most* 5 *edges.* Then, Euler's formula implies that every facet of every cell is a pentagon. This case is excluded by Lemma 11.3.

Let $F_1 = \mathbf{v}_1 \mathbf{v}_2 \ldots \mathbf{v}_6$ be a hexagonal facet of D_1. Since no angle of F_1 can be greater than $2\pi/3$, every angle of F_1 is $2\pi/3$. Without loss of generality, we may assume that $F_1 = D_1 \cap D_2$ and let $\mathbf{x}_{2+i}$, $i = 1, 2, \ldots, 6$, be the points such that $\mathbf{v}_i \mathbf{v}_{i+1}$ is the radical axis of $\mathbf{x}_1$, $\mathbf{x}_2$, and $\mathbf{x}_{i+2}$, where $\mathbf{v}_7 = \mathbf{v}_1$. Then, the six tetrahedra $\mathbf{x}_1 \mathbf{x}_2 \mathbf{x}_3 \mathbf{x}_4$, $\mathbf{x}_1 \mathbf{x}_2 \mathbf{x}_4 \mathbf{x}_5$, $\ldots$, $\mathbf{x}_1 \mathbf{x}_2 \mathbf{x}_8 \mathbf{x}_3$ satisfy the condition of Lemma 11.1, and their circumradii are not greater than τ. Thus all of them are congruent to T^* and, in particular, $\|\mathbf{x}_1, \mathbf{x}_2\| = 2$ and

$$\|\mathbf{x}_1, \mathbf{x}_3\| = \|\mathbf{x}_1, \mathbf{x}_5\| = \|\mathbf{x}_1, \mathbf{x}_7\| = 4/\sqrt{3}. \tag{11.9}$$

For convenience, we denote the midpoint of $\mathbf{x}_1 \mathbf{x}_i$ by $\mathbf{m}_i$. Then, by routine computation it follows that

$$\|\mathbf{m}_2, \mathbf{v}_i\| = \sqrt{2/3}, \quad i = 1, 2, \ldots, 6,$$

and therefore, since every angle of F_1 is $2\pi/3$, F_1 is a regular hexagon. By the observation just above Lemma 11.2, we can draw a regular truncated octahedron G of side length $\sqrt{2/3}$ and centered at $\mathbf{x}_1$, say $\mathbf{x}_1 = \mathbf{o}$, such that F_1 is a facet of G and the facets of G and D_1 adjacent to F_1 lie in the same planes. Without loss of generality, we may assume that the three facets of G along the edges $\mathbf{v}_1 \mathbf{v}_2$, $\mathbf{v}_3 \mathbf{v}_4$, and $\mathbf{v}_5 \mathbf{v}_6$ are squares.

We proceed to show that $D_1 = G$. First, we prove that the three facets F_2, F_4, and F_6 of D_1 that join F_1 at edges $\mathbf{v}_1\mathbf{v}_2$, $\mathbf{v}_3\mathbf{v}_4$, and $\mathbf{v}_5\mathbf{v}_6$, respectively, are squares (see Figure 11.5). By (11.9) and Lemma 11.1, every angle of F_2 is less than $2\pi/3$. This, together with the fact that the angles of F_2 at $\mathbf{v}_1$ and $\mathbf{v}_2$ are $\pi/2$, implies that F_2 is a quadrangle, say $\mathbf{v}_1\mathbf{v}_2\mathbf{w}_2\mathbf{w}_1$. Let $\mathbf{x}_9$ be the point such that the line $\mathbf{w}_1\mathbf{w}_2$ is the radical axis of $\mathbf{x}_1$, $\mathbf{x}_3$, and $\mathbf{x}_9$. Now, we consider the triangle $\mathbf{x}_1\mathbf{x}_3\mathbf{x}_9$. Since $\|\mathbf{x}_1, \mathbf{x}_9\| \geq 2$, $\|\mathbf{x}_3, \mathbf{x}_9\| \geq 2$, and $\|\mathbf{x}_1, \mathbf{x}_3\| = b = 4/\sqrt{3}$, we have $\angle \mathbf{x}_1\mathbf{x}_9\mathbf{x}_3 < \pi/2$. Thus, routine analysis yields that the distance from the center of the circumcircle of $\mathbf{x}_1\mathbf{x}_3\mathbf{x}_9$ to the line $\mathbf{x}_1\mathbf{x}_3$ attains its minimum $\sqrt{1/6}$ when

$$\|\mathbf{x}_1, \mathbf{x}_9\| = \|\mathbf{x}_3, \mathbf{x}_9\| = 2.$$

This means that the line $\mathbf{w}_1\mathbf{w}_2$ does not intersect the open circle

$$C = \left\{\mathbf{x} \in H_2 : \|\mathbf{x}, \mathbf{m}_3\| < \sqrt{1/6}\right\},$$

where H_2 indicates the plane determined by F_2 and $\mathbf{m}_i$ indicates the midpoint of $\mathbf{x}_1\mathbf{x}_i$. On the other hand, the distance between $\mathbf{x}_1$ and any point $\mathbf{x} \in H_2$ that does not belong to the circle

$$C' = \left\{\mathbf{x} \in H_2 : \|\mathbf{x}, \mathbf{m}_3\| \leq \sqrt{1/3}\right\}$$

is greater than τ. Thus, both $\mathbf{w}_1$ and $\mathbf{w}_2$ belong to C'. We observe that C and C' are, in fact, the incircle and circumcircle of the corresponding square facet F_2' of G. Therefore, we have $F_2 = F_2'$.

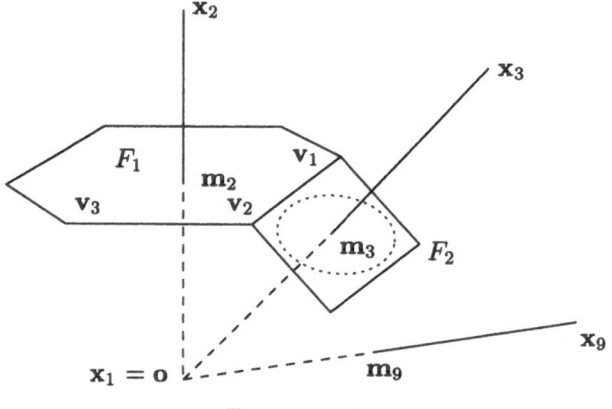

Figure 11.5

Let F_3 be the facet of D_1 joining F_1 at $\mathbf{v}_2\mathbf{v}_3$, and let $\mathbf{v}_2$, $\mathbf{v}_3$, $\mathbf{u}_1$, $\mathbf{u}_2$, $\mathbf{u}_3$, and $\mathbf{w}_2$ be its consecutive vertices. Then, $\mathbf{v}_2$, $\mathbf{w}_2 \in F_2$ and $\mathbf{v}_3$, $\mathbf{u}_1 \in F_4$. By considering the angles at $\mathbf{u}_1$ and $\mathbf{w}_2$, it follows that the edges $\mathbf{u}_1\mathbf{u}_2$ and $\mathbf{u}_3\mathbf{w}_2$ lie on the lines determined by the corresponding edges of G. If

$u_2 = u_3$, routine computation yields that $\|x_1, u_2\| > \tau$. Hence, F_3 is a regular hexagon.

Repeating this process, it follows that $D_1 = G$. Then, by considering the neighbors of D_1, we obtain that $D_i = G + x_i$ and therefore X is a space-centered cubic lattice. Theorem 11.1 follows.   □

Clearly, we have $r(S_n) < 1$ in every dimension. We also have the following lower bound for $r(S_n)$.

Theorem 11.2 (Ryškov and Horváth [1]).

$$r(S_n) \geq \sqrt[n]{\frac{\theta(S_n)}{\delta(S_n)}} - 1 \gg 0.51466.$$

Proof. Let $S_n + X$ be a sphere packing in which the maximal radius of the spherical holes is $r(S_n)$. Then

$$(1 + r(S_n))S_n + X$$

is a covering in E^n. From the definitions of $\delta(S_n)$ and $\theta(S_n)$ it follows that

$$\theta(S_n) \leq \lim_{l \to \infty} \frac{\operatorname{card}\{X \cap lI_n\}v((1 + r(S_n))S_n)}{v(lI_n)}$$

$$= \lim_{l \to \infty} \frac{\operatorname{card}\{X \cap lI_n\}v(S_n)}{v(lI_n)}(1 + r(S_n))^n$$

$$\leq (1 + r(S_n))^n \delta(S_n).$$

In other words,

$$r(S_n) \geq \sqrt[n]{\frac{\theta(S_n)}{\delta(S_n)}} - 1.$$

Also, by Theorem 3.4, Corollary 7.2, and Theorem 8.2 we have

$$r(S_n) \geq \sqrt[n]{\frac{\theta(S_n)}{\delta(S_n)}} - 1 \gg 0.51466.$$

Theorem 11.2 is proved.   □

To end this section, let us list a couple of open problems.

Problem 11.1. *Is it true that for all $n \geq 2$,*

$$r(S_{n+1}) \geq r(S_n)?$$

Problem 11.2. *Determine the value of $\overline{\lim_{n \to \infty}} r(S_n)$.*

11.2. Spherical Holes in Lattice Sphere Packings

Let $S_n + \Lambda$ be a lattice sphere packing in E^n, and write

$$\mu(\Lambda) = r(S_n, \Lambda) + 1. \tag{11.10}$$

Usually, $\mu(\Lambda)$ is called the depth of the *deepest holes* of Λ. Clearly, we have

$$\mu(\Lambda) = \max_{\mathbf{x} \in E^n} \min_{\mathbf{z} \in \Lambda} \{\|\mathbf{x}, \mathbf{z}\|\} = \max_{\mathbf{x} \in D(\mathbf{o})} \{\|\mathbf{o}, \mathbf{x}\|\}. \tag{11.11}$$

In this section we will introduce several results about $r^*(S_n)$ by studying $\mu(\Lambda)$.

Theorem 11.3 (Rogers [3]).

$$r^*(S_n) < 2.$$

Proof. More generally, we prove the following constructive assertion.

Assertion 11.1. *If $S_n + \Lambda$ is a lattice sphere packing with $\mu(\Lambda) \geq 3$, then there exists a new lattice sphere packing $S_n + \Lambda^*$ such that $\Lambda \subset \Lambda^*$ and*

$$\det(\Lambda^*) = \tfrac{1}{3}\det(\Lambda).$$

It follows from (11.11) that there is a point $\mathbf{x} \in E^n$ such that

$$\|\mathbf{x}, \mathbf{z}\| \geq \mu(\Lambda), \quad \mathbf{z} \in \Lambda. \tag{11.12}$$

Also, it is clear that

$$\mu(\tfrac{1}{3}\Lambda) = \tfrac{1}{3}\mu(\Lambda),$$

and therefore there exists a lattice point $\mathbf{u} \in \Lambda$ such that

$$\|\mathbf{x}, \tfrac{1}{3}\mathbf{u}\| \leq \tfrac{1}{3}\mu(\Lambda). \tag{11.13}$$

Clearly, $\tfrac{1}{3}\mathbf{u} \notin \Lambda$ and thus

$$\det(\Lambda^*) = \tfrac{1}{3}\det(\Lambda),$$

where

$$\Lambda^* = \Lambda \bigcup \{\tfrac{1}{3}\mathbf{u} + \Lambda\} \bigcup \{\tfrac{2}{3}\mathbf{u} + \Lambda\}.$$

Then it follows from (11.12), (11.13), and the assumption of Assertion 11.1 that

$$\|\tfrac{1}{3}\mathbf{u}, \mathbf{z}\| \geq \|\mathbf{x}, \mathbf{z}\| - \|\mathbf{x}, \tfrac{1}{3}\mathbf{u}\| \geq \tfrac{2}{3}\mu(\Lambda) \geq 2$$

holds for every point $\mathbf{z} \in \Lambda$, and therefore

$$\|\mathbf{u}^*, \mathbf{v}^*\| \geq 2$$

for any distinct points $\mathbf{u}^\star$ and $\mathbf{v}^\star$ of $\Lambda^\star$. In other words, $S_n + \Lambda^\star$ is a packing.

Clearly, Theorem 11.3 is a consequence of (11.10) and Assertion 11.1. □

Using a very complicated existence method, based upon the work of C.A. Rogers and C.L. Siegel, G.J. Butler was able to improve the previous theorem as follows.

Theorem 11.4 (Butler [1]).

$$r^*(S_n) \leq 1 + o(1).$$

Remark 11.1. *Both Rogers' original result and Butler's improvement dealt not only with spheres, but also with general centrally symmetric convex bodies. By modifying Rogers' constructive method for spheres, Henk [1] obtained*

$$r^*(S_n) \leq \sqrt{21}/2 - 1.$$

Remark 11.2. *Analogously to Theorem 11.2 we have the follwing lower bound for $r^*(S_n)$:*

$$r^*(S_n) \geq \sqrt[n]{\frac{\theta^*(S_n)}{\delta^*(S_n)}} - 1 \gg 0.51466.$$

By studying different patterns of the Dirichlet-Voronoi cells $D(\mathbf{o})$ of the lattice sphere packings in E^4 and E^5, J. Horváth proved the following result.

Theorem 11.5 (Horváth [2]).

$$r^*(S_4) = \sqrt{2\sqrt{3}\left(\sqrt{3} - 1\right)} - 1$$

and

$$r^*(S_5) = \sqrt{\frac{3}{2} + \frac{\sqrt{13}}{6}} - 1.$$

The proofs of both Theorem 11.4 and Theorem 11.5 are very complicated. We omit them here. We list three more problems.

Problem 11.3. *Is it true that for all $n \geq 2$,*

$$r^*(S_{n+1}) \geq r^*(S_n)?$$

Problem 11.4. *Determine the value of $\overline{\lim_{n \to \infty}} r^*(S_n)$.*

Problem 11.5. *Is there a dimension n such that $r^*(S_n) \geq 1$?*

It is conjectured (see Conway and Sloane [1], Gruber and Lekkerkerker [1], and Rogers [14]) that the answer to Problem 11.5 is yes. If so,

$$\delta(S_n) > \delta^*(S_n)$$

holds in the corresponding dimension.

11.3. Cylindrical Holes in Lattice Sphere Packings

In 1960, in a short note A. Heppes proved the following result: *In every three-dimensional lattice sphere packing there is a straight line of infinite length that does not intersect any of the spheres.* In fact, his proof implied the existence of *cylindrical holes* (of infinite length) in every three-dimensional lattice sphere packing. Let $\rho^*(S_n)$ be the maximum number such that in every n-dimensional lattice sphere packing $S_n + \Lambda$ there is a cylindrical hole with an $(n-1)$-dimensional spherical base of radius $\rho^*(S_n)$. Improving Heppes' result, I. Hortobágyi determined the value of $\rho^*(S_3)$. Also, J. Horváth and S.S. Ryškov obtained a lower bound for $\rho^*(S_n)$ in high dimensions. In this section we will introduce the main results of this kind.

For convenience, we denote by $H(\mathbf{x}_1, \mathbf{x}_2, \ldots, \mathbf{x}_m)$ the hyperplane determined by $\mathbf{x}_1, \mathbf{x}_2, \ldots, \mathbf{x}_m$. A *Seeber set* of an n-dimensional lattice Λ is a set of n points $\{\mathbf{u}_1, \mathbf{u}_2, \ldots, \mathbf{u}_n\}$ such that for $i = 1, 2, \ldots, n$, $\mathbf{u}_i$ is a point of $\Lambda \setminus H(\mathbf{o}, \mathbf{u}_{i-1}, \mathbf{u}_{i-2}, \ldots)$ with minimal norm. We now introduce a basic lemma concerning the structure of a lattice.

Lemma 11.5 (Horváth [3]). *Let Λ be a packing lattice of S_n and let $\{\mathbf{u}_1, \mathbf{u}_2, \ldots, \mathbf{u}_n\}$ be a Seeber set of Λ. Denote by H_i the hyperplane $\{\mathbf{x} : \langle \mathbf{x}, \mathbf{u}_i \rangle = 0\}$ and by $p_i(\mathbf{x})$ the projection of $\mathbf{x}$ onto H_i. Then*

$$\|\mathbf{o}, p_i(\mathbf{u})\| \geq \sqrt{3}$$

holds for every point $\mathbf{u} \in \Lambda \setminus H(\mathbf{o}, \mathbf{u}_i)$.

Proof. First we proceed to show that for any point $\mathbf{u} \in \Lambda$,

$$\|\mathbf{o}, \mathbf{u}_i\| \leq \max \{\|\mathbf{o}, \mathbf{u}\|, \|\mathbf{u}, \mathbf{u}_i\|\} . \tag{11.14}$$

If $\mathbf{u} \notin H(\mathbf{o}, \ldots, \mathbf{u}_{i-1})$, then from the definition of a Seeber set it follows that

$$\|\mathbf{o}, \mathbf{u}_i\| \leq \|\mathbf{o}, \mathbf{u}\|.$$

If $\mathbf{u} \in H(\mathbf{o}, \ldots, \mathbf{u}_{i-1})$, then $\mathbf{u} - \mathbf{u}_i \notin H(\mathbf{o}, \ldots, \mathbf{u}_{i-1})$ and hence

$$\|\mathbf{o}, \mathbf{u}_i\| \leq \|\mathbf{u}, \mathbf{u}_i\|.$$

These two cases imply (11.14).

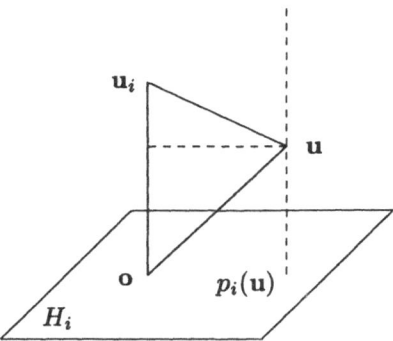

Figure 11.6

Now, assuming $\mathbf{u} \in \Lambda \setminus H(\mathbf{o}, \mathbf{u}_i)$, we consider the triangle $\mathbf{ouu}_i$. It follows from (11.14) that $\mathbf{ou}_i$ is not the longest side of the triangle, and therefore (see Figure 11.6)

$$\max\{\angle \mathbf{ou}_i\mathbf{u}, \angle \mathbf{uou}_i\} \geq \pi/3.$$

On the other hand, since $p_i(\mathbf{u}) = p_i(\mathbf{u} + k\mathbf{u}_i)$ for any integer k, we may assume that

$$\max\{\angle \mathbf{ou}_i\mathbf{u}, \angle \mathbf{uou}_i\} \leq \pi/2.$$

Thus, since $\|\mathbf{o}, \mathbf{u}\| \geq 2$ and $\|\mathbf{u}, \mathbf{u}_i\| \geq 2$,

$$\|\mathbf{o}, p_i(\mathbf{u})\| \geq 2\sin(\pi/3) = \sqrt{3}.$$

Lemma 11.5 is proved.   □

Although it is simple, this lemma plays an important role in the work of Heppes, Hortobágyi, Horváth, and Ryškov on the cylindrical holes in lattice sphere packings.

Theorem 11.6 (Heppes [2] and Hortobágyi [1]).

$$\rho^*(S_3) = \frac{3\sqrt{2}}{4} - 1.$$

In other words, every lattice sphere packing $S_3 + \Lambda$ has a cylindrical hole with a circular base of radius $3\sqrt{2}/4 - 1$.

Proof. Let Λ be a packing lattice of S_3, and let $\{\mathbf{u}_1, \mathbf{u}_2, \mathbf{u}_3\}$ be a Seeber set of Λ. We aim to show that if $rS_2 + p_1(\Lambda)$ is a covering in H_1, then $r \geq 3\sqrt{2}/4$. In other words, letting $T = \mathbf{v}_1\mathbf{v}_2\mathbf{v}_3$ (where $\mathbf{v}_1 = \mathbf{o}$) be a triangle with $\mathbf{v}_i \in p_1(\Lambda)$ and letting $r(T)$ be its circumradius, we proceed to show that

$$r(T) \geq \frac{3\sqrt{2}}{4}. \tag{11.15}$$

For convenience, we denote the circumcircle of T by S, and denote the cylinder with base S by Q.

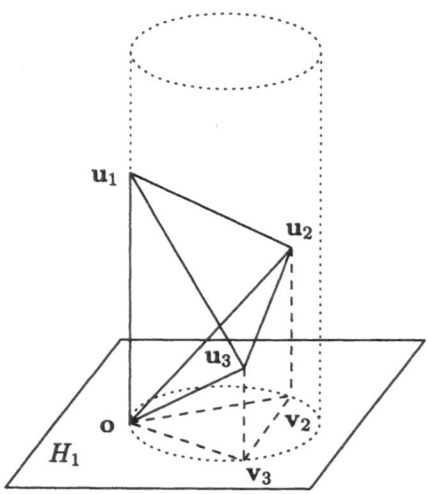

Figure 11.7

By Lemma 11.5, every side of T has length not less than $\sqrt{3}$. Thus, simple computation yields that $r(T) \geq 3\sqrt{2}/4$ if one of its sides is greater than 2. So, without loss of generality, we may assume that $\|\mathbf{o}, \mathbf{u}_1\| = 2$,

$$\mathbf{v}_i = p_1(\mathbf{u}_i), \quad i = 1, \, 2, \, 3,$$

$$\sqrt{3} \leq \|\mathbf{v}_i, \mathbf{v}_j\| \leq 2, \quad i \neq j,$$

and that the dihedral angle of $\mathbf{o}\mathbf{u}_1\mathbf{u}_2\mathbf{u}_3$ at the edge $\mathbf{o}\mathbf{u}_3$ is not greater than $\pi/2$. We deal with two cases (see Figure 11.7).

Case 1. *Both $\mathbf{u}_2$ and $\mathbf{u}_3$ lie on the same side of the plane $H_1 + \frac{1}{2}\mathbf{u}_1$.* Without loss of generality, we may assume that $\mathbf{o}$, $\mathbf{u}_2$, and $\mathbf{u}_3$ lie on the same side of the plane $H_1 + \frac{1}{2}\mathbf{u}_1$ and that $\mathbf{u}_3$ is the nearest to it. Then, we adjust the positions of $\mathbf{u}_2$ and $\mathbf{u}_3$ using the following process: **Step 1.** Replace $\mathbf{u}_3$ by its projection onto the plane $H_1 + \frac{1}{2}\mathbf{u}_1$. **Step 2.** Move $\mathbf{u}_3$ along the boundary of $S + \frac{1}{2}\mathbf{u}_1$ towards $\frac{1}{2}\mathbf{u}_1$ until

$$\|\mathbf{u}_1, \mathbf{u}_3\| = \|\mathbf{o}, \mathbf{u}_3\| = 2.$$

Step 3. Move $\mathbf{u}_2$ along the line parallel with $\mathbf{o}\mathbf{u}_1$ until we have $\|\mathbf{u}_2, \mathbf{u}_3\| = 2$. **Step 4.** Move $\mathbf{u}_2$ along the curve $(\mathrm{bd}(2S_3) + \mathbf{u}_3) \cap \mathrm{bd}(Q)$ towards the plane $H_1 + \frac{1}{2}\mathbf{u}_1$ until we have

$$\|\mathbf{o}, \mathbf{u}_2\| = \|\mathbf{u}_2, \mathbf{u}_3\| = 2.$$

Step 5. Move $\mathbf{u}_2$ along the circle $\{\mathbf{x} : \langle \mathbf{x} - \frac{1}{2}\mathbf{u}_3, \mathbf{u}_3 \rangle = 0, \|\mathbf{x}, \frac{1}{2}\mathbf{u}_3\| = \sqrt{3}\}$ in the direction from $\mathbf{u}_2$ to $\mathbf{u}_1$ so that $\|\mathbf{u}_1, \mathbf{u}_2\| = 2$. Then we obtain a regular tetrahedron $\mathbf{o}\mathbf{u}_1\mathbf{u}_2\mathbf{u}_3$. Simple analysis shows that in this process the corresponding lattices are packing lattices of S_3 and the corresponding circumradii $r(T)$ do not increase. Thus, (11.15) is true in this case.

Case 2. *The plane $H_1 + \frac{1}{2}\mathbf{u}_1$ separates $\mathbf{u}_2$ and $\mathbf{u}_3$.* Without loss of generality, we may assume that $\mathbf{o}$ and $\mathbf{u}_2$ lie on the same side of $H_1 + \frac{1}{2}\mathbf{u}_1$. Then, we adjust the positions of $\mathbf{u}_2$ and $\mathbf{u}_3$ as follows: **Step 1.** Move $\mathbf{u}_2$ along the line parallel with $\mathbf{o}\mathbf{u}_1$ towards H_1 until $\|\mathbf{o}, \mathbf{u}_2\| = 2$. Then move $\mathbf{u}_3$ similarly until $\|\mathbf{u}_2, \mathbf{u}_3\| = 2$. **Step 2.** Move $\mathbf{u}_3$ along the curve $\big(\mathrm{bd}(2S_3) + \mathbf{u}_2\big) \cap \mathrm{bd}(Q)$ towards the plane $H_1 + \frac{1}{2}\mathbf{u}_1$ so that $\|\mathbf{u}_1, \mathbf{u}_3\| = 2$. Then, we have

$$\|\mathbf{o}, \mathbf{u}_1\| = \|\mathbf{o}, \mathbf{u}_2\| = \|\mathbf{u}_2, \mathbf{u}_3\| = \|\mathbf{u}_1, \mathbf{u}_3\| = 2,$$

$\|\mathbf{o}, \mathbf{u}_3\| > 2$, and $\|\mathbf{u}_1, \mathbf{u}_3\| > 2$. **Step 3.** Move $\mathbf{u}_2$ along the circle $\{\mathbf{x} : \langle \mathbf{x} - \frac{1}{2}\mathbf{u}_3, \mathbf{u}_3 \rangle = 0, \|\mathbf{x}, \frac{1}{2}\mathbf{u}_3\| = c\}$, where c is a constant, in the direction from $\mathbf{u}_2$ to $\mathbf{u}_1$ so that $\|\mathbf{u}_1, \mathbf{u}_2\| = 2$. Then, adjust the position of $\mathbf{u}_3$ accordingly. At the end of this process we obtain a regular tetrahedron $\mathbf{o}\mathbf{u}_1\mathbf{u}_2\mathbf{u}_3$. Just as in Case 1, in this process the corresponding lattices are packing lattices of S_3 and the corresponding circumradii $r(T)$ do not increase. Therefore, (11.15) is true in this case.

Hence, Theorem 11.6 is proved.   □

Remark 11.3. *By looking at the proof more closely, it seems that the extreme lattice sphere packing is unique up to isometry.*

Between $\rho^*(S_n)$ and $r^*(S_n)$ we have the following relationship, which provides a lower bound for $\rho^*(S_n)$.

Theorem 11.7 (Ryškov and Horváth [1]).

$$\rho^*(S_n) \geq \frac{\sqrt{3}(1 + r^*(S_{n-1}))}{2} - 1 \gg 0.3117.$$

Proof. Let $S_n + \Lambda$ be a lattice sphere packing with the thinnest cylinder hole, and let $\{\mathbf{u}_1, \mathbf{u}_2, \ldots, \mathbf{u}_n\}$ be a Seeber set of Λ. Then, it follows from Lemma 11.5 that

$$(\sqrt{3}/2)S_{n-1} + p_i(\Lambda)$$

is a lattice sphere packing in H_i. From the definition of $r^*(S_{n-1})$ we have

$$r \geq \frac{\sqrt{3}(1 + r^*(S_{n-1}))}{2}$$

if $rS_{n-1} + p_i(\Lambda)$ is a covering in H_i. On the other hand, we have

$$p_i(S_n + \mathbf{u}) = S_{n-1} + p_i(\mathbf{u})$$

for every point $\mathbf{u} \in \Lambda$. Thus, using Remark 11.2,

$$\rho^*(S_n) \geq \frac{\sqrt{3}(1 + r^*(S_{n-1}))}{2} - 1 \gg 0.3117.$$

Theorem 11.7 is proved. □

Remark 11.4. *From the proof of Theorem 11.7 it follows that every lattice sphere packing $S_n + \Lambda$ has cylindrical holes, with spherical bases of radii at least $\sqrt{3}(1 + r^*(S_{n-1}))/2 - 1$, in n independent directions.*

Analogous to Problems 11.2 and 11.4, we have the following problem concerning $\rho^*(S_n)$.

Problem 11.6. *Determine the value of $\overline{\lim\limits_{n \to \infty}} \rho^*(S_n)$.*

12. Problems of Blocking Light Rays

12.1. Introduction

We call a straight line with one end extending to infinity a *light ray*. Let $h(S_n)$ be the smallest number such that there exists a finite sphere packing $S_n + X$, with $X = \{o, x_1, x_2, \ldots, x_{h(S_n)}\}$, such that every light ray starting from o is blocked by one of the translates $S_n + x_i$, $x_i \in X \setminus \{o\}$. Similarly, when X is restricted to being a subset of a lattice, we denote the corresponding number by $h^*(S_n)$. Hornich proposed the following problem (see L. Fejes Tóth [7]).

Hornich's Problem: *Determine the numbers $h(S_n)$ and $h^*(S_n)$.*

In 1959, L. Fejes Tóth [7] obtained the first result in this direction, proving

$$h(S_3) \geq 19.$$

He also proposed another related problem. Let $\ell(S_n)$ be the smallest number such that there is a finite packing $S_n + Y$, where $Y = \{o, y_1, y_2, \ldots, y_{\ell(S_n)}\}$, and such that every light ray starting from any point of S_n is blocked by one of the translates $S_n + y_j$, $y_j \in Y \setminus \{o\}$. His problem can be formulated as follows.

L. Fejes Tóth's Problem: *Determine the number $\ell(S_n)$.*

Remark 12.1. *Let $\ell^*(S_n)$ be the corresponding number in the lattice case. It follows from Theorem 11.6 and Theorem 11.7 that when $n \geq 3$,*

$$\ell^*(S_n) = \infty.$$

It is obvious that

$$h(S_n) \le h^*(S_n), \tag{12.1}$$

$$h(S_n) \le \ell(S_n), \tag{12.2}$$

and, in particular,

$$h(S_2) = h^*(S_2) = \ell(S_2) = \ell^*(S_2) = 6.$$

Also, relating $k(S_n)$ and $h(S_n)$ we have the following simple result.

Theorem 12.1. *If $n \ge 3$, then*

$$k(S_n) < h(S_n).$$

Proof. Let S_n, $S_n + \mathbf{x}_1, \ldots, S_n + \mathbf{x}_{k(S_n)}$ be an optimal kissing configuration, and let S_n, $S_n + \mathbf{y}_1, \ldots, S_n + \mathbf{y}_{h(S_n)}$ be a finite packing such that every light ray starting from $\mathbf{o}$ is blocked by one of the translates $S_n + \mathbf{y}_j$, $j = 1$, $2, \ldots, h(S_n)$.

For any point $\mathbf{x}$, let $\Omega(\mathbf{x})$ be the intersection of $\mathrm{bd}(S_n)$ and the smallest cone with vertex $\mathbf{o}$, containing $S_n + \mathbf{x}$. Clearly, $\Omega(\mathbf{x})$ is a cap on $\mathrm{bd}(S_n)$. For convenience, we denote the surface area of a cap of geodesic radius $\pi/6$ by μ_n. From the definitions of $k(S_n)$ and $h(S_n)$ it follows that the caps $\Omega(\mathbf{x}_i)$, $i = 1, 2, \ldots, k(S_n)$, form a packing in $\mathrm{bd}(S_n)$, the caps $\Omega(\mathbf{y}_j)$, $j = 1, 2$, $\ldots, h(S_n)$, form a covering in $\mathrm{bd}(S_n)$,

$$s(\Omega(\mathbf{x}_i)) = \mu_n,$$

$$s(\Omega(\mathbf{y}_j)) \le \mu_n,$$

and therefore

$$k(S_n)\mu_n = \sum_{i=1}^{k(S_n)} s(\Omega(\mathbf{x}_i)) < n\omega_n$$
$$< \sum_{j=1}^{h(S_n)} s(\Omega(\mathbf{y}_j))$$
$$\le h(S_n)\mu_n.$$

Theorem 12.1 follows.   □

Remark 12.2. *Halberg, Levin, and Straus [1] and Boltjanski and Gohberg [1] studied an analogous problem by allowing overlap between the translates $S_n + \mathbf{x}_i$, $\mathbf{x}_i \in X \setminus \{\mathbf{o}\}$. It follows from routine argument that this case is equivalent to a cap covering problem. Namely, what is the minimal number of caps of geodesic radii $\pi/6$ that can cover the surface of S_n?*

Remark 12.3. *Substituting S_n by any centrally symmetric convex body C in Hornich's problem and by any convex body K in L. Fejes Tóth's problem, one can define and study $h(C)$, $h^*(C)$, $\ell(K)$, and $\ell^*(K)$. We call these numbers the Hornich number, lattice Hornich number, L. Fejes Tóth number, and lattice L. Fejes Tóth number of the corresponding convex body, respectively. Clearly, these numbers are invariant under linear transformations of the considered convex bodies.*

12.2. Hornich's Problem

Clearly, determining the values of $h(S_n)$ and $h^*(S_n)$ for $n \geq 3$ is extraordinary difficult. Perhaps, it is even more difficult than determining the values of $k(S_n)$ and $k^*(S_n)$. In E^3, L. Fejes Tóth's lower bound for $h(S_3)$ was improved by Heppes [3] and Csóka [1], while Danzer [1] obtained an upper bound. The known results about $h(S_3)$ can be expressed as

$$30 \leq h(S_3) \leq 42.$$

Since these results are still far from best possible and their methods are very technical, we will not go into the details. In this section we will obtain general bounds for $h(S_n)$ and $h^*(S_n)$. First, let us introduce a geometric version of *Dirichlet's approximation theorem*.

Lemma 12.1 (Henk [2]). *Let C be an n-dimensional centrally symmetric convex body centered at $\mathbf{o}$, and let Λ be a lattice such that $C+\Lambda$ is a packing. For any vector $\mathbf{v} \neq \mathbf{o}$ there is a point $\mathbf{u} \in \Lambda$ and a positive integer*

$$j \leq \frac{2^n \delta(C)}{\delta(C, \Lambda)}$$

such that

$$j\mathbf{v} \in \text{int}(C) + \mathbf{u}.$$

Proof. Writing

$$m = \left\lfloor \frac{2^n \delta(C)}{\delta(C, \Lambda)} \right\rfloor,$$

we consider the set

$$X = \bigcup_{i=0}^{m} (\Lambda + i\mathbf{v}).$$

If $\frac{1}{2}C + X$ is a packing, then

$$\delta(\tfrac{1}{2}C, X) = (m+1)\delta(\tfrac{1}{2}C, \Lambda)$$

$$= \frac{(m+1)\delta(C,\Lambda)}{2^n}$$

$$> \frac{2^n\delta(C)}{\delta(C,\Lambda)}\frac{\delta(C,\Lambda)}{2^n}$$

$$= \delta(C),$$

which is impossible. Therefore,

$$\text{int}\left(\tfrac{1}{2}C + i_1\mathbf{v} + \mathbf{u}_1\right) \cap \text{int}\left(\tfrac{1}{2}C + i_2\mathbf{v} + \mathbf{u}_2\right) \neq \emptyset$$

holds for certain $0 \le i_1 < i_2 \le m$, $\mathbf{u}_1 \in \Lambda$, and $\mathbf{u}_2 \in \Lambda$. In other words,

$$(i_2 - i_1)\mathbf{v} \in \text{int}(C) + \mathbf{u}_1 - \mathbf{u}_2.$$

Thus, taking $j = i_2 - i_1$ and $\mathbf{u} = \mathbf{u}_1 - \mathbf{u}_2$, Lemma 12.1 is proved. $\qquad\square$

Theorem 12.2 (Zong [8]).

$$h(C) \le h^*(C) \le \frac{\delta(C)^n 2^{n^2(1+o(1))}}{\delta^*(C)^n},$$

$$h(S_n) \le h^*(S_n) \le 2^{1.401n^2(1+o(1))}.$$

Proof. It is well-known from convex geometry (see John [1] or Gruber and Lekkerkerker [1]) that for any n-dimensional centrally symmetric convex body C there is a linear transformation L such that

$$S_n \subseteq L(C) \subseteq \sqrt{n}S_n.$$

Thus, without loss of generality, we may assume that

$$S_n \subseteq C \subseteq \sqrt{n}S_n. \tag{12.3}$$

Let $C + \Lambda$ be a lattice packing with density $\delta^*(C)$. By Lemma 12.1, for every vector $\mathbf{v}$ with $\|\mathbf{v}\| = \sqrt{n} + 1$ there is a positive integer

$$j \le \frac{2^n\delta(C)}{\delta^*(C)}$$

and a point $\mathbf{u} \in \Lambda \setminus \{\mathbf{o}\}$ such that

$$j\mathbf{v} \in \text{int}(C) + \mathbf{u}. \tag{12.4}$$

Let D be a Dirichlet-Voronoi cell associated with Λ. By Assertion 11.1, Remark 11.1, and (12.3) we have

$$d(D) \le 3(\sqrt{n} + 1). \tag{12.5}$$

Thus, writing

$$m = \left\lfloor \frac{2^n \delta(C)}{\delta^*(C)} \right\rfloor,$$

by (12.3), (12.4), and (12.5) we have

$$h^*(C) \le \text{card}\left\{ m(\sqrt{n}+1)S_n \cap \Lambda \right\}$$
$$\le \frac{\omega_n \left(m(\sqrt{n}+1) + d(D_i) \right)^n \delta^*(C)}{v(C)}$$
$$\le \frac{\omega_n \left(m(\sqrt{n}+1) + 3(\sqrt{n}+1) \right)^n \delta^*(C)}{\omega_n}$$
$$\le \frac{\delta(C)^n 2^{n^2(1+o(1))}}{\delta^*(C)^n}.$$

This proves the first part of Theorem 12.2. Then, taking $C = S_n$, the second part follows from Theorem 3.1 and Theorem 8.2. □

Remark 12.4. *The original result of Zong [8], which is based on the number-theoretic version of Dirichlet's approximation theorem, is weaker than Theorem 12.2.*

As a counterpart to Theorem 12.2 we have the following lower bound for $h(S_n)$.

Theorem 12.3 (Bárány and Leader [1]).

$$h(S_n) \ge 2^{0.275n^2(1+o(1))}.$$

Proof. For convenience we write $r = 2^{0.275n}$, $\mu = 1/n$, and $r_i = 2 + i\mu$. Let X be a subset of rS_n such that $S_n + X \cup \{o\}$ is a finite packing, and write

$$X_i = \left\{ \mathbf{x} \in X : r_i \le \|\mathbf{x}\| < r_{i+1} \right\}.$$

Clearly we have $X_i = \emptyset$ if $i > nr$, and

$$X = \bigcup_{i=0}^{nr} X_i. \tag{12.6}$$

Suppose that $\mathbf{x} \in X_i$, and denote by $\varphi/2$ the spherical radius of the cap $(S_n + \mathbf{x}) \cap \text{bd}(r_i S_n)$. Then (see Figure 12.1), for large n,

$$\sin \frac{\varphi}{2} = \sqrt{1 - \cos^2 \frac{\varphi}{2}} \ge \sqrt{1 - \left(\frac{r_i^2 + r_{i+1}^2 - 1}{2r_i r_{i+1}} \right)^2}$$
$$\ge \frac{1}{r_i} \sqrt{\frac{15}{16} \left(\frac{n-4}{n} \right)}.$$

Thus, by (8.1) and (8.21) we have

$$\begin{aligned}
\operatorname{card}\{X_i\} \le m(n, \varphi/2) &= m[n, \varphi] \\
&\le \left(2^{0.599} \sin(\varphi/2)\right)^{-n} \\
&\le r^n 2^{-0.552n(1+o(1))}.
\end{aligned} \tag{12.7}$$

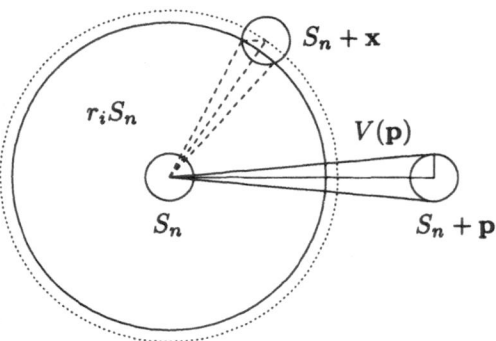

Figure 12.1

Let $\mathbf{p}$ be an arbitrary point in E^n. Let $V(\mathbf{p})$ be the cone with vertex $\mathbf{o}$ over $S_n + \mathbf{p}$, denote by $\Omega(\mathbf{p})$ the cap $V(\mathbf{p}) \cap \operatorname{bd}(S_n)$, and let β be the spherical radius of $\Omega(\mathbf{p})$. Then,

$$\begin{aligned}
\frac{s(\Omega(\mathbf{p}))}{s(S_n)} &\le \frac{(n-1)\omega_{n-1}}{n\omega_n} \int_0^\beta (\sin\theta)^{n-2} d\theta \\
&\le \frac{\omega_{n-1}}{n\omega_n} (\sin\beta)^{n-1} \le n\|\mathbf{p}\|^{1-n}.
\end{aligned} \tag{12.8}$$

By (12.6), (12.7), and (12.8) we have

$$\begin{aligned}
\frac{\sum_{\mathbf{x} \in X} s(\Omega(\mathbf{x}))}{s(S_n)} &= \sum_{i=0}^{nr} \sum_{\mathbf{x} \in X_i} \frac{s(\Omega(\mathbf{x}))}{n\omega_n} \\
&\le 2^{-0.552n(1+o(1))} \sum_{i=0}^{nr} r_i \\
&\le r^2 2^{-0.552n(1+o(1))} \\
&= 2^{-0.002n(1+o(1))}.
\end{aligned}$$

This means that $S_n + X$ can block at most half of the light rays starting from $\mathbf{o}$ when n is large. Therefore, considering the light rays not blocked by $S_n + X$ and using (12.8) we have

$$h(S_n) \ge \frac{r^{n-1}}{2n} \ge 2^{0.275n^2(1+o(1))}.$$

Theorem 12.3 is proved. □

To end this section, we propose the following problem.

Problem 12.1. *Do there exist absolute constants c and c^* such that*

$$h(S_n) = 2^{cn^2(1+o(1))} \quad or \quad h^*(S_n) = 2^{c^*n^2(1+o(1))}?$$

If so, determine them.

12.3. L. Fejes Tóth's Problem

From the physical point of view, L. Fejes Tóth's problem is more natural, more fascinating, and more challenging than Hornich's problem. In 1996, Böröczky and Soltan [1] proved that

$$\ell(S_n) < \infty,$$

which shows the fundamental difference between $\ell(S_n)$ and $\ell^*(S_n)$ (see Remark 12.1). Almost simultaneously, Zong [6] obtained the following upper bound for $\ell(S_n)$ using a constructive method.

Theorem 12.4 (Zong [6]).

$$\ell(S_n) \leq (8e)^n(n+1)^{n-1}n^{(n^2+n-2)/2}.$$

Proof. For convenience, we consider $S = (\sqrt{n}/2)S_n$ instead of S_n and assume that $n \geq 3$. Clearly, we have

$$I_n \subset S \subset \sqrt{n}I_n. \tag{12.9}$$

Let α be a positive number strictly less than one, and take

$$k = \left\lfloor \frac{\sqrt{n}}{\alpha} + 1 \right\rfloor. \tag{12.10}$$

Then,

$$\alpha I_n \subset S \subset k\alpha I_n.$$

Denote the intersections of αI_n and $k\alpha I_n$ with the hyperplane $\{\mathbf{x} \in E^n : x_i = 0\}$ by $J(i)$ and $J'(i)$, respectively. Let $X_i = \{\mathbf{x}_1, \mathbf{x}_2, \ldots, \mathbf{x}_{k^{n-1}}\}$ be a set of points such that $J(i) + X_i$ is a tiling in $J'(i)$, and write

$$Y_i = \left\{\mathbf{x}_j + j\sqrt{n}\mathbf{e}_i : \ \mathbf{x}_j \in X_i\right\}, \tag{12.11}$$

where $\mathbf{e}_i$ indicates the ith unit basis vector. Then, it follows from (12.9) and (12.11) that $S + Y_i \cup \{\mathbf{o}\}$ is a finite packing. Let $\mathbf{w}$ be a point with $w_i > 0$. We introduce another set of points

$$Z_i(\mathbf{w}) = \left\{ \frac{w_i + j\sqrt{n}}{w_i}(\mathbf{w} + \mathbf{x}_j) : \ \mathbf{x}_j \in X_i \right\}. \tag{12.12}$$

Let $\mathbf{x}'_j$, $\mathbf{y}'_j$, and $\mathbf{z}'_j$ be three geometrically corresponding points in $X_i + \mathbf{w}$, $Y_i + \mathbf{w}$, and $Z_i(\mathbf{w})$, respectively. Then, $\mathbf{z}'_j$ is the intersection of the hyperplane $\{\mathbf{x} \in E^n : x_i = y'_{ji}\}$ and the light ray of direction $\mathbf{x}'_j$. By (12.9) and (12.12) it is clear that $S + Z_i(\mathbf{w}) \cup \{\mathbf{o}\}$ is another finite packing.

First, let us prove the following statement.

Assertion 12.1. *When $\mathbf{w}$ is a point with $w_i \geq 3k^{n-1}n/2(1 - \alpha)$, the spheres $S + Z_i(\mathbf{w})$ block all the light rays starting from S and passing through $J'(i) + \mathbf{w}$.*

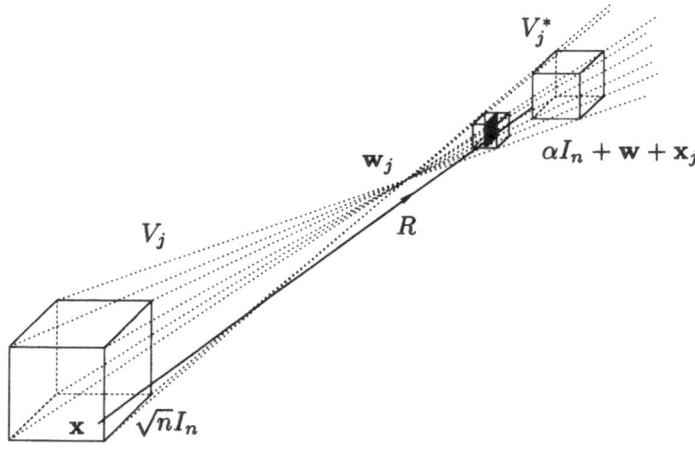

Figure 12.2

Let $\mathbf{w}$ be a point with

$$w_i \geq \frac{k^{n-1}\sqrt{n}(\sqrt{n} + \alpha)}{1 - \alpha}, \tag{12.13}$$

and let $\mathbf{x}_j$ be a point of X_i. Take

$$\mathbf{w}_j = \frac{\sqrt{n}}{\alpha + \sqrt{n}}(\mathbf{w} + \mathbf{x}_j)$$

and denote by V_j, V_j^* the two cones with the same vertex $\mathbf{w}_j$ and over $\sqrt{n}I_n$, $\alpha I_n + \mathbf{w} + \mathbf{x}_j$, respectively. Then (see Figure 12.2), we have

$$V_j^* - \mathbf{w}_j = -V_j + \mathbf{w}_j \tag{12.14}$$

and

$$V_j^* = \bigcup_{\beta \geq 0} \left(\frac{\beta + \sqrt{n}}{\alpha + \sqrt{n}} (\mathbf{w} + \mathbf{x}_j) + \beta I_n \right). \tag{12.15}$$

If R is a light ray starting from a point $\mathbf{x} \in V_j$ and passing through a point $\mathbf{y} \in V_j^*$, $\mathbf{x} \neq \mathbf{y}$, then the rest of the light ray is contained in V_j^*. Otherwise, let H be the two-dimensional plane containing both R and $\mathbf{w}_j$. By considering the intersection $H \cap (V_j \cup V_j^*)$ and applying the symmetry indicated by (12.14) one can easily obtain a contradiction. On the other hand, using (12.13) and since $j \leq k^{n-1}$, it is easy to get $\beta \leq 1$ when

$$\frac{\beta + \sqrt{n}}{\alpha + \sqrt{n}} = \frac{w_i + j\sqrt{n}}{w_i}.$$

Since $J(i) \subset \alpha I_n$, by (12.9) and (12.14) the light rays starting from $\sqrt{n} I_n$ and passing through $J(i) + \mathbf{w} + \mathbf{x}_j$ can be blocked by the translate

$$S + \frac{w_i + j\sqrt{n}}{w_i} (\mathbf{w} + \mathbf{x}_j).$$

Then, Assertion 12.1 follows from the fact that $J(i) + X_i + \mathbf{w}$ is a tiling in $J'(i) + \mathbf{w}$.

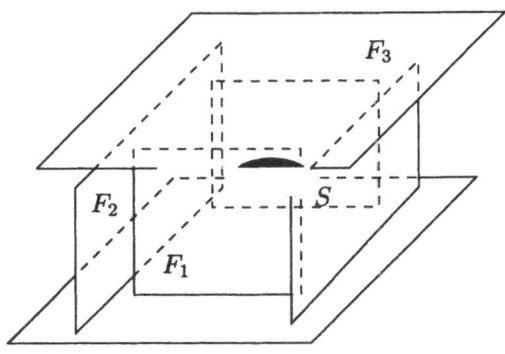

Figure 12.3

Write

$$P(\xi_1, \xi_2, \ldots, \xi_n) = \left\{ \mathbf{x} \in E^n : |x_i| \leq \tfrac{1}{2}\xi_i, \ i = 1, \ 2, \ldots, n \right\}$$

and denote its facet with $x_i = \xi_i/2$ by $F_i(\xi_1, \xi_2, \ldots, \xi_n)$. Take $l = k\alpha \geq \sqrt{n}$ and

$$m = 2 \left\lfloor \frac{3k^{n-1}n}{2(1-\alpha)} + 1 \right\rfloor. \tag{12.16}$$

Abbreviate $F_1(ml, ml, \ldots, ml)$, $F_2(2ml, (m + k^{n-1})l, ml, \ldots, ml)$, $\ldots$, $F(2ml, 2ml, \ldots, 2ml, (m+k^{n-1})l)$ by $F_1, F_2, \ldots, F_n$, respectively. Clearly, we have

$$\frac{ml}{2} \geq \frac{3k^{n-1}n}{2(1-\alpha)}. \tag{12.17}$$

We proceed to prove the following assertion.

Assertion 12.2. *Every light ray starting from S intersects at least one of the $2n$ facets $F_1, -F_1, F_2, -F_2, \ldots, F_n, -F_n$ (see Figure 12.3).*

For convenience, we write $F_i^* = F_i(l, l, \ldots, l)$. Then F_i blocks all the light rays starting from lI_n and passing through a fixed point $\mathbf{p} \in \mathrm{bd}(mlI_n)$ if and only if F_i blocks all the light rays starting from F_i^* and passing through $\mathbf{p}$, in other words, as showed in Figure 12.4, if and only if

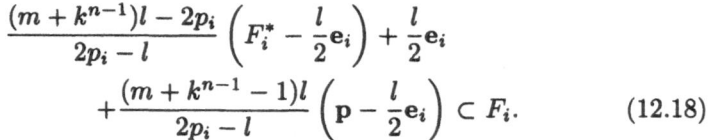

$$\frac{(m + k^{n-1})l - 2p_i}{2p_i - l}\left(F_i^* - \frac{l}{2}\mathbf{e}_i\right) + \frac{l}{2}\mathbf{e}_i$$
$$+ \frac{(m + k^{n-1} - 1)l}{2p_i - l}\left(\mathbf{p} - \frac{l}{2}\mathbf{e}_i\right) \subset F_i. \tag{12.18}$$

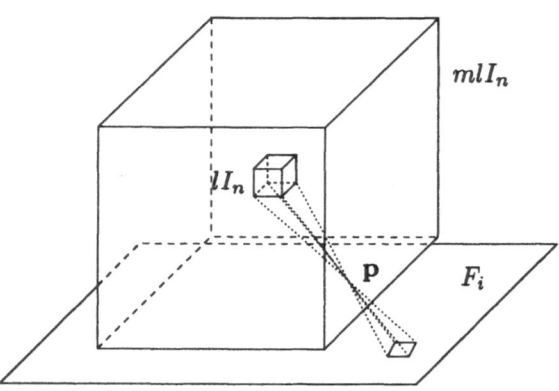

Figure 12.4

Define q_i reductively by

$$q_n = \frac{l}{2}\left(\frac{m}{2m-1}(m + k^{n-1} - 1) + 1\right),$$
$$q_i = \frac{m + k^{n-1} - 1}{m - 1}q_{i+1} + \frac{l}{2}$$

for $i = n - 1, n - 2, \ldots, 2$, and $q_1 = ml/2$. Then, by (12.10), (12.16), and the assumption $n \geq 3$,

$$\frac{(m + k^{n-1} + 1)l}{4} \leq q_n \leq q_{n-1} \leq \cdots \leq q_2$$

$$\leq \left(\frac{m + k^{n-1} - 1}{m - 1}\right)^{n-2} \left(q_n + \frac{(n-2)l}{2}\right)$$

$$\leq \left(1 + \frac{1}{2n}\right)^{n-2} \left(\frac{1}{2} + \frac{1}{3n}\right)\frac{ml}{2}$$

$$\leq \frac{1}{2}\left(1 + \frac{1}{2n}\right)^{n} \frac{1 + 2/3n}{(1 + 1/2n)^2}\frac{ml}{2}$$

$$\leq \frac{\sqrt{e}}{2}\frac{ml}{2} < \frac{ml}{2}. \tag{12.19}$$

From (12.18), (12.19), and routine computation based upon elementary geometry illustrated by Figure 12.4, it can be shown that every light ray starting from F_i^* and passing through a point $\mathbf{p} = (p_1, p_2, \ldots, p_n) \in$ bd(mlI_n) with

$$p_i \geq \max_{i+1 \leq j \leq n} \left\{\frac{m + k^{n-1} - 1}{m - 1}q_j + \frac{l}{2}\right\} = q_i$$

and

$$|p_j| \leq q_j, \quad j = i + 1, \ldots, n,$$

can be blocked by F_i. On the other hand, it follows from (12.19) that for each point $\mathbf{p} \in$ bd(mlI_n) there is an index i such that

$$|p_i| \geq q_i \quad \text{and} \quad |p_j| \leq q_j$$

for $j = i + 1, \ldots, n$. Therefore, by (12.9) all the light rays starting from S can be blocked by the union of the $2n$ facets $F_1, -F_1, F_2, -F_2, \ldots, F_n, -F_n$. Assertion 12.2 follows.

Let $W_i = \{\mathbf{w}_{i,1}, \mathbf{w}_{i,2}, \ldots, \mathbf{w}_{i,\tau_i}\}$ be a set of points such that $J'(i) + W_i$ is a tiling in F_i, where

$$\tau_i = 2^{i-1}m^{n-1}. \tag{12.20}$$

We define

$$\mathfrak{S}_i = \bigcup_{\mathbf{w} \in W_i} Z_i(\mathbf{w})$$

and

$$\mathfrak{S} = \bigcup_{i=1}^{n} \left\{\mathfrak{S}_i \bigcup (-\mathfrak{S}_i)\right\}.$$

Based on the periodic distribution of $Y_i + W_i$ and the relationship between $Y_i + W_i$ and $\mathfrak{S}_i$, it is easy to see that

$$\text{int}(S + \mathbf{v}_1) \cap \text{int}(S + \mathbf{v}_2) = \emptyset$$

whenever $\mathbf{v}_1$ and $\mathbf{v}_2$ are different points in $\mathfrak{S}_i$. Also, from the definition of F_i and by looking at the $(j + t)$th coordinates of $\mathbf{v}_j$ and $\mathbf{v}_{j+t}$, it can be verified that

$$\text{int}(S + \mathbf{v}_j) \cap \text{int}(S + \mathbf{v}_{j+t}) = \emptyset$$

whenever $\mathbf{v}_j \in \Im_j$, $\mathbf{v}_{j+t} \in \Im_{j+t}$, and t is a positive integer. Therefore, $S + \Im \cup \{\mathbf{o}\}$ is a finite packing. On the other hand, by Assertion 12.1, Assertion 12.2, and (12.17) the spheres $S + \Im$ block all the light rays starting from S. Hence, by (12.10), (12.16), and (12.20),

$$\ell(S_n) = \ell(S) \le \text{card}\{\Im\} = 2 \sum_{i=1}^{n} \text{card}\{\Im_i\}$$

$$= 2k^{n-1} \sum_{i=1}^{n} \tau_i = 2\left(2^n - 1\right)(km)^{n-1}$$

$$\le 2^n \left(2^n - 1\right) \left(\frac{3k^n n}{2(1-\alpha)} + k\right)^{n-1}$$

$$\le 8^n \left(\frac{n^{(n+2)/2}}{\alpha^n(1-\alpha)}\right)^{n-1}$$

holds for every number α with $0 < \alpha < 1$. Since the function

$$g(\alpha) = \alpha^n(1 - \alpha)$$

attains its maximum $\frac{1}{n+1}\left(\frac{n}{n+1}\right)^n$ at $\alpha = \frac{n}{n+1}$, we have

$$\ell(S_n) \le 8^n \left(1 + \frac{1}{n}\right)^{n^2} (n+1)^{n-1} n^{(n^2+n-2)/2}$$

$$\le (8e)^n (n+1)^{n-1} n^{(n^2+n-2)/2}.$$

Theorem 12.4 is proved.   □

Remark 12.5. *As a generalization of John's result, Leichtweiß* [1] *proved that for every n-dimensional convex body K there is a linear transformation L such that*

$$S_n \subseteq L(K) \subseteq nS_n$$

(for a proof of this result, we refer to Assertion 13.2). Applying Leichtweiß and John's results, Zong [6] *proved, in a similar manner to Theorem 12.4, that*

$$\ell(K) \le (8e)^n (n+1)^{n-1} n^{3(n^2-1)/2}$$

for every n-dimensional convex body K, and that

$$\ell(C) \le (8e)^n (n+1)^{n-1} n^{n^2-1}$$

for every n-dimensional centrally symmetric convex body C.

Improving the upper bound for $\ell(C)$ with a nonconstructive method, we have the following theorem.

Theorem 12.5.
$$\ell(C) \le 48^{n^2(1+o(1))}.$$

Remark 12.6. *Using a modification of Zong's method, Talata [3] obtained a similar bound for general convex bodies. Meanwhile Bárány discovered a proof for the sphere case based on the ideas of Section 2. Our proof here is based on his method.*

In order to prove Theorem 12.5, as well as Henk's lemma (Lemma 12.1), we require another basic result.

Lemma 12.2 (Rogers and Zong [1]). *Let K_1 and K_2 be two convex bodies in E^n. Then K_2 can be covered by $\lfloor v(K_1 - K_2)\theta(K_2)/v(K_1)\rfloor$ translates of K_1. In particular, when K_1 is centrally symmetric and $K_2 = \gamma K_1$, K_2 can be covered by $\lfloor (1+\gamma)^n \theta(K_1)\rfloor$ translates of K_1.*

Proof. For convenience, we assume that both K_1 and K_2 contain o. Let X be a set of points such that $K_1 + X$ is a covering in E^n with covering density $\theta(K_1)$, and let l be a large number. For any point $\mathbf{y} \in E^n$ we define

$$f(\mathbf{y}) = \text{card}\left\{\mathbf{x} \in X : (K_2 + \mathbf{y}) \cap (K_1 + \mathbf{x}) \ne \emptyset\right\}.$$

It is easy to show that

$$(K_2 + \mathbf{y}) \cap (K_1 + \mathbf{x}) \ne \emptyset$$

if and only if

$$\mathbf{y} - \mathbf{x} \in K_1 - K_2.$$

Thus, letting $\chi(\mathbf{y})$ be the characteristic function of $K_1 - K_2$, we have

$$\int_{lI_n} f(\mathbf{y})d\mathbf{y} = \sum_{\mathbf{x} \in X} \int_{lI_n} \chi(\mathbf{y} - \mathbf{x})d\mathbf{y}$$
$$\le v(K_1 - K_2)\,\text{card}\left\{X \cap (l + 2d(K_1))\,I_n\right\}$$
$$\le v(K_1 - K_2)\frac{v((l + 2d(K_1))I_n)(\theta(K_1) + \epsilon)}{v(K_1)}.$$

Therefore,

$$f(\mathbf{y}) \le \left\lfloor \frac{v(K_1 - K_2)}{v(K_1)}\theta(K_1) \right\rfloor$$

holds for some point $\mathbf{y} \in E^n$. The general case is proved. In the symmetric case, it is easy to see that

$$K_1 - K_2 = (1 + \gamma)K_1.$$

Thus, the assertion follows from the general case.   □

Proof of Theorem 12.5. Assume that C is centered at $\mathbf{o}$, and let Λ be a lattice such that $C + \Lambda$ is a packing with density $\delta^*(C)$. For convenience, we write $r = \frac{1}{4}$,

$$m = \left\lfloor \frac{\delta(C)2^n}{r^n \delta^*(C)} \right\rfloor,$$
$$m' = \lfloor 3^n \theta(C) \rfloor,$$
$$r_i = 14(m' + i - 1)r,$$

and

$$L(i,j) = \{\mathbf{u} \in \Lambda : (2rC + \mathbf{u}) \cap \mathrm{bd}(jr_iC) \neq \emptyset\}.$$

Clearly, we have

$$2rC + L(i,j) \subset (jr_i + 4r)C \setminus (jr_i - 4r)C. \tag{12.21}$$

By Lemma 12.1, for every point $\mathbf{v} \in \mathrm{bd}(r_iC)$ there is a $j \leq m$ and a lattice point $\mathbf{u} \in \Lambda$ such that

$$j\mathbf{v} \in rC + \mathbf{u}$$

and therefore (see Figure 12.5)

$$rC + j\mathbf{v} \subset 2rC + \mathbf{u}.$$

This means that all the light rays starting from rC can be blocked by

$$\bigcup_{j=1}^{m} (2rC + L(i,j)).$$

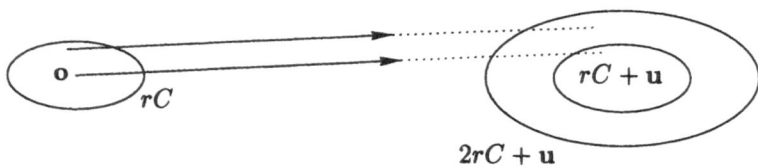

$$2rC + \mathbf{u}$$

Figure 12.5

By Lemma 12.2, we assume that $2rC$ can be covered by m' translates $rC + \mathbf{x}_1, rC + \mathbf{x}_2, \ldots, rC + \mathbf{x}_{m'}$. Then we have

$$\mathbf{x}_i \in 3rC. \tag{12.22}$$

Defining

$$L'(i,j) = \left\{ \frac{jm' + i - 1}{j(m' + i - 1)} \mathbf{u} : \mathbf{u} \in L(i,j) \right\} \tag{12.23}$$

and
$$L^\star(i,j) = L'(i,j) + \mathbf{x}_i, \tag{12.24}$$
it follows that all the light rays starting from $rC + \mathbf{x}_i$ can be blocked by
$$\bigcup_{j=1}^{m}(2rC + L^\star(i,j)).$$

Therefore, all the light rays starting from $2rC$ can be blocked by
$$\bigcup_{i=1}^{m'}\bigcup_{j=1}^{m}(2rC + L^\star(i,j)). \tag{12.25}$$

Since $C + \Lambda$ is a packing and
$$\frac{1}{2} \le \frac{jm' + i - 1}{j(m' + i - 1)} \le 1,$$

$2rC + L'(i,j)$ is a finite packing of $2rC$. Also, by (12.21), (12.22), (12.23), and (12.24) it follows that
$$2rC + L^\star(i,j) \subset (jr_1 + 7(2i-1)r)C \setminus (jr_1 + 7(2i-3)r)C.$$

Denoting the set on the right-hand side of the above formula by $C(i,j)$, it can be verified that the $\mathrm{int}(C(i,j))$ are pairwise disjoint. Thus, the system (12.25) is a finite packing of $2rC$ in $(m+1)r_1C$. Hence, by Theorem 3.1 and Remark 3.5,
$$\ell(C) \le \frac{v((m+1)r_1C)}{v(2rC)} \le \frac{(m+1)^n r_1^n}{2^n r^n}$$
$$\le 48^{n^2(1+o(1))}.$$

Theorem 12.5 is proved. $\square$

Based on (12.2), Theorem 12.3, and Theorem 12.5 we propose the following problems to end this section.

Problem 12.2. *Does there exist an absolute constant c such that*
$$\ell(S_n) = 2^{cn^2(1+o(1))}?$$

If so, determine it.

Problem 12.3. *Let $S_n + X$ be a packing in E^n, let $\varsigma(S_n, X)$ be the maximal length of the segments contained in $E^n \setminus \{\mathrm{int}(S_n) + X\}$, and let*
$$\varsigma(S_n) = \min_X \varsigma(S_n, X).$$

Does there exist a constant c such that

$$\varsigma(S_n) = 2^{cn(1+o(1))}?$$

12.4. László Fejes Tóth

László Fejes Tóth was born in 1915 in Szeged, Hungary. In 1933 he entered the University of Budapest to study mathematics and physics, where L. Fejér was one of his teachers. In 1938 he obtained a doctorate in mathematics.

Soon afterwards, L. Fejes Tóth served in the Hungarian army for two years. In 1941 he left the army and became an assistant at the University of Kolozsvár. Four years later, he moved to Budapest and worked as a middle-school teacher there. At the same time he lectured on mathematics at the University of Budapest. In 1949 he became a professor of mathematics at the Technical University of Veszprém. In 1965, Professor L. Fejes Tóth moved to the Mathematical Institute of the Hungarian Academy of Sciences, where he served as the director from 1970 to 1985. He retired in 1985. Professor L. Fejes Tóth was a guest of many foreign universities such as the University of Freiburg, University of Wisconsin, University of Washington, Ohio State University, and the University of Salzburg.

Professor L. Fejes Tóth mainly worked in the areas of convex and discrete geometry, in which he published about 180 papers and two books. He has never had a Ph.D student. However, his work and open problems have influenced almost every contemporary discrete geometer, both domestic and foreign.

Professor L. Fejes Tóth is a member of the Hungarian Academy of Sciences and a corresponding member of the Sächsische Akademie der Wissenschaften zu Leipzig. He has been awarded honorary doctorates by the University of Salzburg and Veszprém University. In addition, he was honored by several Hungarian national prizes and a Gauss medal of the Braunschweigische Wissenschaftliche Gesellschaft.

13. Finite Sphere Packings

13.1. Introduction

How can one pack m unit spheres in E^n such that the volume or the surface area of their convex hull is minimal?

Let m be a positive integer, and let $\mathbf{u}$ be a unit vector in E^n. Then, we define

$$L_m = \{2i\mathbf{u} : \ i = 0,\ 1,\ \ldots,\ m-1\}$$

and

$$S_{m,n} = \operatorname{conv}\{S_n + L_m\}.$$

Clearly, $S_n + L_m$ is a packing in $S_{m,n}$. Writing

$$\mu_m(S_n) = \frac{mv(S_n)}{v(S_{m,n})},$$

it can be regarded as a local density of $S_n + L_m$ in $S_{m,n}$. By routine computation it follows that

$$\mu_m(S_n) = \frac{m\omega_n}{\omega_n + 2(m-1)\omega_{n-1}} \gg \delta(S_n).$$

Based on this observation, in 1975 L. Fejes Tóth [10] made the following conjecture about the volume case of the above problem.

Sausage Conjecture: *In E^n, $n \geq 5$, the volume of the convex hull of m nonoverlapping unit spheres is at least*

$$\omega_n + 2(m-1)\omega_{n-1}.$$

Also, equality is attained only if their centers are equally spaced on a line, a distance 2 apart.

On the other hand, for the surface area case, Croft, Falconer, and Guy [1] formulated the following counterpart.

Spherical Conjecture: *When packing m unit spheres into a convex set so that the surface area is minimal, it seems likely that the optimum shape is roughly spherical if m is large.*

From an intuitive point of view, it is hard to imagine that both conjectures are correct. Nevertheless, the sausage conjecture was proved by Betke, Henk, and Wills [1] for sufficiently large n, while the spherical conjecture was confirmed independently by Böröczky Jr. [2] and Zong [1]. In this chapter we will deal mainly with the work of U. Betke, K. Böröczky Jr., M. Henk, J.M. Wills, and C. Zong.

13.2. The Spherical Conjecture

The following theorem provides a positive solution for the spherical conjecture.

Theorem 13.1 (Böröczky Jr. [2] and Zong [1]). *Let $\mathcal{K}^{\circ}_{m,n}$ be the family of convex bodies $K^{\circ}_{m,n}$ that contain m nonoverlapping n-dimensional unit spheres and at which the surface area attains its minimum. Write*

$$\eta_m = \max_{K^{\circ}_{m,n} \in \mathcal{K}^{\circ}_{m,n}} \min_{\substack{r S_n + x \subseteq K^{\circ}_{m,n} \subseteq r' S_n + x \\ x \in E^n}} \frac{r'}{r}.$$

Then

$$\lim_{m \to \infty} \eta_m = 1.$$

To prove this theorem, in addition to Theorem 1.2 and the *isoperimetric inequality* we need the following lemma.

Lemma 13.1. *For every n-dimensional convex body K there is a rectangular parallelepiped P such that*

$$P \subseteq K \subseteq n^{3/2} P.$$

Proof. First, let us quote two fundamental results from Leichtweiß [1]. These results are not only useful in the proof of this theorem, but are also interesting and important.

Assertion 13.1. *Let α and β be two fixed numbers with $-1 \leq \alpha \leq 1$ and $-1 \leq \beta \leq 1$. If among all the ellipsoids*

$$E_\lambda = \left\{ \mathbf{x} \in E^n : \lambda(x_1 - \alpha)(x_1 + \beta) + \sum_{i=1}^n (x_i)^2 \leq 1 \right\},$$

where $\lambda \geq 0$, the unit sphere takes the minimum volume, then $\alpha\beta \geq 1/n$.

It follows from routine computation that

$$v(E_\lambda) = \left(1 + (1 + \alpha\beta)\lambda + \frac{(\alpha + \beta)^2}{4}\lambda^2 \right)^{n/2} (1 + \lambda)^{-(n+1)/2} \omega_n$$

and therefore

$$\left. \frac{\partial(v(E_\lambda))}{\partial \lambda} \right|_{\lambda=0} = \frac{n\alpha\beta - 1}{2} \omega_n.$$

Thus, the assumption $v(E_\lambda) \geq \omega_n$ for all $\lambda \geq 0$ implies

$$\left. \frac{\partial(v(E_\lambda))}{\partial \lambda} \right|_{\lambda=0} \geq 0,$$

and consequently

$$\alpha\beta \geq \tfrac{1}{n}.$$

Assertion 13.1 follows.

Assertion 13.2. *For every n-dimensional convex body K there exists an ellipsoid E such that*

$$\tfrac{1}{n}E \subseteq K \subseteq E.$$

Assume that E is a minimal circumscribed ellipsoid of K; for convenience we take E to be centered at o. Let L be a linear transformation of E^n such that $L(E) = S_n$. Then, S_n is a minimal circumscribed ellipsoid of $L(K)$. If

$$H_1 = \{\mathbf{x} \in E^n : x_1 = \alpha\}$$

and

$$H_2 = \{\mathbf{x} \in E^n : x_1 = -\beta\},$$

where $0 \leq \alpha \leq 1$ and $0 \leq \beta \leq 1$ are two supporting hyperplanes of $L(K)$, then every point $\mathbf{x} \in L(K)$ satisfies both

$$\sum_{i=1}^n (x_i)^2 \leq 1$$

and

$$(x_1 - \alpha)(x_1 + \beta) \leq 0.$$

Thus, it follows that

$$L(K) \subseteq E_\lambda$$

for all $\lambda \geq 0$ and therefore ω_n is the minimal value of $v(E_\lambda)$. Then, by Assertion 13.1 we have $1/n \leq \alpha \leq 1$,

$$\tfrac{1}{n} S_n \subseteq L(K) \subseteq S_n,$$

and finally

$$\tfrac{1}{n} E \subseteq K \subseteq E.$$

This proves Assertion 13.2.

By Assertion 13.2, if $\mathbf{a}_1, \mathbf{a}_2, \ldots, \mathbf{a}_n$ are the n axes of E with lengths $\varrho_1, \varrho_2, \ldots, \varrho_n$, respectively, and if

$$P = \left\{ \mathbf{x} \in E^n : \langle \mathbf{x}, \mathbf{a}_i \rangle \leq n^{-3/2} \varrho_i, \; i = 1, 2, \ldots, n \right\},$$

then we have

$$P \subseteq \tfrac{1}{n} E \subseteq K \subseteq E \subseteq n^{3/2} P.$$

Lemma 13.1 is proved.   □

Proof of Theorem 13.1. By Lemma 13.1, for every convex body $K_{m,n}^\circ$ there is a rectangular parallelepiped P such that

$$P \subseteq K_{m,n}^\circ \subseteq n^{3/2} P. \tag{13.1}$$

Let $l_1, l_2, \ldots, l_n$ be the lengths of the n edges of P. It follows from (13.1) that

$$m\omega_n \leq v(K_{m,n}^\circ) \leq v(n^{3/2} P) = n^{3n/2} \prod_{i=1}^{n} l_i \tag{13.2}$$

and

$$s(K_{m,n}^\circ) \geq s(P) = \prod_{i=1}^{n} l_i \sum_{j=1}^{n} \frac{2}{l_j}. \tag{13.3}$$

On the other hand, by Theorem 1.2, m unit spheres can be packed into a large sphere of radius

$$r = 2 \sqrt[n]{\frac{m}{\delta(S_n)}}$$

when m is sufficiently large. Then, by (13.2), (13.3), and the minimum assumption of $s(K_{m,n}^\circ)$ we have

$$s(P) \leq s(K_{m,n}^\circ) \leq s(rS_n),$$

$$\prod_{i=1}^{n} l_i \sum_{j=1}^{n} \frac{2}{l_j} \le n\omega_n \left(2\sqrt[n]{\frac{m_+}{\delta(S_n)}} \right)^{n-1}$$

$$\le n\omega_n \left(2\sqrt[n]{\frac{v(K_{m,n}^{\circ})}{\delta(S_n)\omega_m}} \right)^{n-1}$$

$$\le c_1(n) \left(\prod_{i=1}^{n} l_i \right)^{(n-1)/n}.$$

This implies

$$\left(\prod_{i=1}^{n} l_i \right)^{1/n} \sum_{i=1}^{n} \frac{1}{l_i} \le c_2(n)$$

and therefore

$$c_3(n) \le \frac{\max_{1 \le i \le n} \{l_i\}}{\min_{1 \le i \le n} \{l_i\}} \le c_4(n), \tag{13.4}$$

where $c_1(n)$, $c_2(n)$, $c_3(n)$, and $c_4(n)$ are all positive numbers depending only on n. Thus, by (13.1) and (13.4), for every convex body $K_{m,n}^{\circ}$ there is a suitable number $\gamma(K_{m,n}^{\circ})$ such that

$$c_5(n)S_n \subseteq \gamma(K_{m,n}^{\circ})K_{m,n}^{\circ} \subseteq S_n,$$

where $c_5(n)$ is another constant depending only upon n. Clearly,

$$\lim_{m \to \infty} \gamma(K_{m,n}^{\circ}) = 0. \tag{13.5}$$

If, contrary to the assertion of Theorem 13.1,

$$\lim_{m \to \infty} \eta_m \ne 1,$$

then by *Blaschke's selection theorem*, there is a nonspherical convex body K and a series of convex bodies $K_{m_1,n}^{\circ}$, $K_{m_2,n}^{\circ}$, ... such that

$$c_5(n)S_n \subseteq K \subseteq S_n$$

and

$$\lim_{m_j \to \infty} \gamma(K_{m_j,n}^{\circ})K_{m_j,n}^{\circ} = K.$$

Then, by the isoperimetric inequality, there are three positive numbers r, α, and β such that

$$v(rS_n) \ge v(K) + \alpha \tag{13.6}$$

and

$$s(rS_n) \le s(K) - \beta. \tag{13.7}$$

Applying Theorem 1.2 once more, it follows from (13.5), (13.6), and (13.7) that $(r/\gamma(K_{m_j,n}^{\circ}))S_n$ contains m_j nonoverlapping unit spheres but with surface area smaller than $s(K^\circ_{m_j,n})$ when m_j is sufficiently large. This contradiction proves

$$\lim_{m\to\infty} \eta_m = 1.$$

Theorem 13.1 is proved.   □

Remark 13.1. *In fact, Theorem 13.1 is true not only for the sphere and surface area but also for any n-dimensional convex body and any fixed quermassintegral between 1 and n (see Böröczky Jr. [2], Zong [1], or Zong [4]).*

13.3. The Sausage Conjecture

For the sausage conjecture we have the following theorem.

Theorem 13.2 (Betke, Henk, and Wills [1]). *Let $K_{m,n}$ be an n-dimensional convex body that contains m nonoverlapping unit spheres. When n is sufficiently large,*

$$v(K_{m,n}) \geq \omega_n + 2(m-1)\omega_{n-1},$$

where equality holds if and only if $K_{m,n}$ is congruent to $S_{m,n}$.

Although the assertion of this theorem is clear and elegant, its proof is long and complicated. Let $S_n + \mathbf{x}_i$, $i = 1, 2, \ldots, m$, be the m unit spheres packed in $K_{m,n}$ and write

$$X_{m,n} = \{\mathbf{x}_1, \mathbf{x}_2, \ldots, \mathbf{x}_m\}.$$

As usual, we denote the Dirichlet-Voronoi cell with respect to $\mathbf{x}_i$ by $D(\mathbf{x}_i)$. Clearly, we have

$$v(K_{m,n}) = \sum_{i=1}^{m} v(D(\mathbf{x}_i) \cap K_{m,n}). \qquad (13.8)$$

Thus, to prove Theorem 13.2 it is sufficient to show that

$$v(D(\mathbf{x}_i) \cap K_{m,n}) \geq 2\omega_{n-1}$$

for $m - 2$ points,

$$v(D(\mathbf{x}_i) \cap K_{m,n}) \geq \omega_{n-1} + \tfrac{1}{2}\omega_n$$

for 2 points, and that equality in these equations cannot hold simultaneously if $K_{m,n}$ is not congruent to $S_{m,n}$. For this purpose we consider a fixed Dirichlet-Voronoi cell, say $D = D(\mathbf{x}_m)$ with respect to $\mathbf{x}_m = \mathbf{o}$, and the corresponding convex body $D^* = D \cap K_{m,n}$.

Definition 13.1. *Let* $\mathbf{u}_i$ *be the unit vector in the direction of* $\mathbf{x}_i$, *and let*

$$\phi = \max_{1 \leq i,\, j \leq m-1} \{\arccos(|\langle \mathbf{u}_i, \mathbf{u}_j \rangle|)\}.$$

Also, without loss of generality, let $\mathbf{u}_1$ *and* $\mathbf{u}_2$ *be two vectors such that*

$$\arccos(|\langle \mathbf{u}_1, \mathbf{u}_2 \rangle|) = \begin{cases} \phi, & \text{if } \phi \geq \pi/3, \\ \phi, & \text{if } \langle \mathbf{u}_i, \mathbf{u}_j \rangle \geq 0 \text{ for all indices}, \\ \max_{\langle \mathbf{u}_i, \mathbf{u}_j \rangle < 0} \{\arccos(|\langle \mathbf{u}_i, \mathbf{u}_j \rangle|)\}, & \text{otherwise.} \end{cases} \quad (13.9)$$

Remark 13.2. *In the third case of* (13.9), *writing*

$$\Omega_1 = \left\{ \mathbf{x} \in \mathrm{bd}(S_n) : \langle \mathbf{u}_1, \mathbf{x} \rangle > \tfrac{1}{2} \right\}$$

and

$$\Omega_2 = \left\{ \mathbf{x} \in \mathrm{bd}(S_n) : \langle \mathbf{u}_1, \mathbf{x} \rangle \leq \langle \mathbf{u}_1, \mathbf{u}_2 \rangle \right\},$$

we have

$$\{\mathbf{u}_1, \mathbf{u}_2, \ldots, \mathbf{u}_{m-1}\} \subset \Omega_1 \cup \Omega_2,$$

and the angle between any two vectors $\mathbf{u} \in \Omega_1$ *and* $\mathbf{v} \in \Omega_2$ *is larger than* $\pi/3$. *By considering two subcases, routine argument shows that*

$$- \cos(\phi/2) \leq \langle \mathbf{u}_1, \mathbf{u}_2 \rangle \leq - \cos \phi. \quad (13.10)$$

Remark 13.3. *Roughly speaking, a small* ϕ *indicates that near to* $\mathbf{o}$ *the convex hull,* $\mathrm{conv}\{S_n + X_{m,n}\}$, *is "sausage-like" if* $\langle \mathbf{u}_1, \mathbf{u}_2 \rangle < 0$ *and is "needle-like" if* $\langle \mathbf{u}_1, \mathbf{u}_2 \rangle > 0$. *Meanwhile, since the sum of the three angles of any triangle is* π, *it follows that* $\phi < \pi/3$ *and* $\langle \mathbf{u}_1, \mathbf{u}_2 \rangle > 0$ *cannot hold at three points of* $X_{m,n}$. *This fact plays a very important role in the proof of Theorem 13.2.*

Definition 13.2. *Write* $G = \mathrm{conv}\{\mathbf{o}, 2\mathbf{u}_1, 2\mathbf{u}_2\} \cap S_n$, *and denote by* $f(G, \mathbf{x})$ *the map given by*

$$f(G, \mathbf{x}) = \mathbf{y} \in G,$$

where

$$\|\mathbf{x}, \mathbf{y}\| = \min_{\mathbf{w} \in G} \{\|\mathbf{x}, \mathbf{w}\|\}.$$

Then, we define

$$\begin{aligned} D_1 &= \{\mathbf{x} \in D^* : f(G, \mathbf{x}) \in \mathrm{rint}(G)\}, \\ D_2 &= \{\mathbf{x} \in D^* : f(G, \mathbf{x}) \in (\mathbf{o}, \mathbf{u}_1) \cup (\mathbf{o}, \mathbf{u}_2)\}, \\ D_3 &= \{\mathbf{x} \in D^* : f(G, \mathbf{x}) = \mathbf{o}\}, \end{aligned}$$

and

$$D_4 = \{\mathbf{x} \in D^* : f(G, \mathbf{x}) \in [2\mathbf{u}_1, 2\mathbf{u}_2]\}.$$

It is clear that

$$v(D^*) \geq \sum_{i=1}^{4} v(D_i). \qquad (13.11)$$

By (13.8) and (13.11), the proof of Theorem 13.2 depends on complicated estimates of the values of $v(D_i)$.

Lemma 13.2. *Let* **u** *and* **v** *be two orthogonal unit vectors, and let* α *and* ϵ *be two positive numbers such that* $\alpha(\alpha + \epsilon) \geq 1$ *and* $(\alpha + \epsilon)\mathbf{v} \in D^*$. *Then*

$$[\mathbf{o}, h_1(\alpha, \epsilon)\mathbf{u}] + \alpha\mathbf{v} \in D^*,$$

where $h_1(\alpha, \epsilon) = \epsilon/\sqrt{(\alpha + \epsilon)^2 - 1}$.

Proof. It follows from the assumption and the convexity of D^* that

$$\text{conv}\{S_n, (\alpha + \epsilon)\mathbf{v}\} \subset D^*. \qquad (13.12)$$

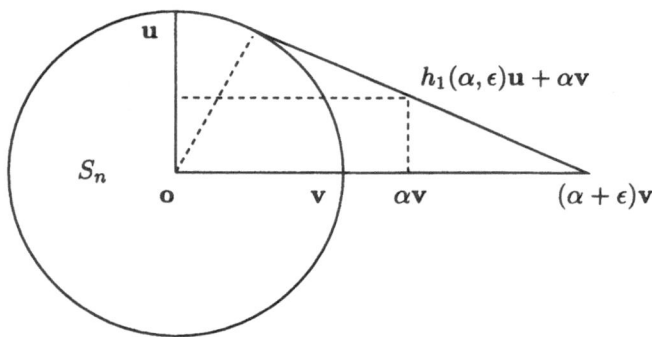

Figure 13.1

With the help of Figure 13.1, since $\alpha \geq 1/(\alpha + \epsilon)$, routine argument yields

$$h_1(\alpha, \epsilon)\mathbf{u} + \alpha\mathbf{v} \in \text{conv}\{S_n, (\alpha + \epsilon)\mathbf{v}\},$$

which implies

$$[\mathbf{o}, h_1(\alpha, \epsilon)\mathbf{u}] + \alpha\mathbf{v} \subset \text{conv}\{S_n, (\alpha + \epsilon)\mathbf{v}\}. \qquad (13.13)$$

Thus, Lemma 13.2 follows from (13.12) and (13.13).   □

Lemma 13.3.
$$v(G) \geq \phi/2.$$

Proof. Write $\gamma = \langle \mathbf{u}_1, \mathbf{u}_2 \rangle$, $\theta = \arccos(|\gamma|)$ and

$$V = \{\alpha\mathbf{u}_1 + \beta\mathbf{u}_2 : \ \alpha \geq 0, \ \beta \geq 0\}.$$

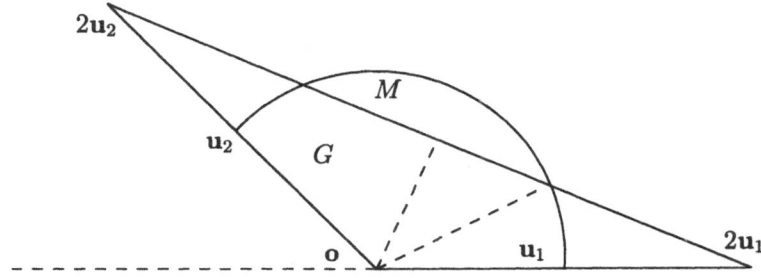

Figure 13.2

Now, as illustrated in Figure 13.2, we deal with two cases.

Case 1. $\gamma \geq -\frac{1}{2}$. Then, we have $\theta = \phi$, $V \cap S_n \subset G$, and therefore

$$v(G) \geq \phi/2. \tag{13.14}$$

Case 2. $\gamma < -\frac{1}{2}$. Writing $M = V \cap S_n \setminus G$, routine computation yields that

$$v(G) = v(V \cap S_n) - v(M)$$
$$= \frac{\pi - \theta}{2} - \arccos x + x\sqrt{1 - x^2},$$

where $x = 2\sin(\theta/2)$. Thus,

$$v(G) - \theta = \frac{\pi}{2} - \arccos x - \frac{3\theta}{2} + x\sqrt{1 - x^2}$$
$$= \arcsin x - 3\arcsin \frac{x}{2} + x\sqrt{1 - x^2}$$
$$\geq 0.$$

Then, by (13.10) it follows that

$$v(G) \geq \theta \geq \phi/2. \tag{13.15}$$

From (13.14) and (13.15) Lemma 13.3 is proved. $\qquad\qquad\square$

Before estimating the values of $v(D_i)$ we introduce some new notation.

$$H = \{\mathbf{x} = \alpha \mathbf{u}_1 + \beta \mathbf{u}_2 : -\infty \leq \alpha, \beta \leq +\infty\},$$
$$H^* = \{\mathbf{x} \in E^n : \langle \mathbf{u}_1, \mathbf{x} \rangle = \langle \mathbf{u}_2, \mathbf{x} \rangle = 0\},$$
$$H_i = \{\mathbf{x} \in E^n : \langle \mathbf{u}_i, \mathbf{x} \rangle = 0\}.$$

We now proceed to estimate the values of $v(D_i)$.

Lemma 13.4.
$$v(D_1) \geq \frac{h_1(1, \operatorname{cosec} \phi - 1)^2 \phi}{2} \omega_{n-2}.$$

Proof. It follows from the definition of ϕ that

$$|\langle \mathbf{u}_i, \mathbf{u}_j \rangle| \geq \cos \phi$$

for $i = 1, 2$ and $j = 1, 2, \ldots, m-1$, which implies

$$\langle \mathbf{u}_j, \mathbf{v} \rangle \leq \sin \phi$$

whenever $\mathbf{v} \in H^* \cap \operatorname{bd}(S_n)$. Hence, from the definition of D^* we obtain

$$\operatorname{cosec} \phi \, (H_i \cap S_n) \subset D^* \tag{13.16}$$

for $i = 1, 2$, and

$$\operatorname{cosec} \phi \, (H^* \cap S_n) \subset D^*.$$

Then, applying Lemma 13.2 with

$$\mathbf{u} \in H \cap \operatorname{bd}(S_n), \quad \mathbf{v} \in H^* \cap \operatorname{bd}(S_n),$$

$\alpha = 1$, and $\epsilon = \operatorname{cosec} \phi - 1$, we have

$$h_1(1, \operatorname{cosec} \phi - 1)G + H^* \cap S_n \subset D_1$$

and therefore, applying Lemma 13.3,

$$v(D_1) \geq \frac{h_1(1, \operatorname{cosec} \phi - 1)^2 \phi}{2} \omega_{n-2}.$$

Lemma 13.4 is proved.   □

Lemma 13.5.
$$v(D_2) \geq h_1(1, \operatorname{cosec} \phi - 1)\omega_{n-1}.$$

Proof. In a similar way to the proof of Lemma 13.4, it follows from (13.16) and Lemma 13.2 that

$$[\mathbf{o}, h_1(1, \operatorname{cosec} \phi - 1)\mathbf{u}_i] + H_i \cap S_n \subset D^* \tag{13.17}$$

for $i = 1, 2$. For convenience, we denote the convex body on the left-hand side of (13.17) by J_i. Let $\mathbf{v}_1$ and $\mathbf{v}_2$ be unit vectors of H determined by $\langle \mathbf{u}_i, \mathbf{v}_i \rangle = 0$ and $\langle \mathbf{u}_i, \mathbf{v}_j \rangle < 0$ (see Figure 13.3), where $\{i, j\} = \{1, 2\}$. Then

$$\{\mathbf{x} \in E^n : \langle \mathbf{v}_i, \mathbf{x} \rangle \geq 0, \ \mathbf{x} \in \operatorname{int}(J_i)\} \subseteq D_2. \tag{13.18}$$

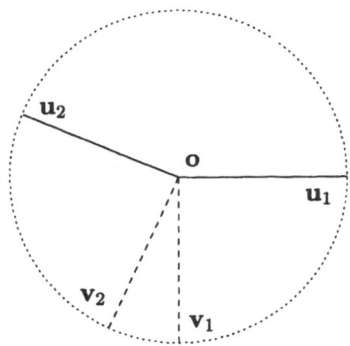

Figure 13.3

Hence, it follows from (13.18) and the definitions of J_i and D_2 that

$$v(D_2) \geq h_1(1, \operatorname{cosec} \phi - 1)\omega_{n-1}.$$

Lemma 13.5 is proved.   □

Lemma 13.6. *If $\phi < \pi/3$ and $\langle u_1, u_2 \rangle > 0$, then*

$$v(D_3) \geq \frac{\pi - \phi}{2\pi}\omega_n.$$

Proof. Writing

$$V^* = \{x = \alpha u_1 + \beta u_2 : \ \alpha \leq 0, \ \beta \leq 0, \ x \in S_n\},$$

routine argument shows that

$$v(V^*) = \frac{\pi - \phi}{2}$$

and

$$(V^* + H^* \cap S_n) \cap S_n \subset D_3.$$

Thus, we have

$$v(D_3) \geq \frac{v(V^*)}{v(S_2)}v(S_n) = \frac{\pi - \phi}{2\pi}\omega_n,$$

which proves our lemma.   □

Lemma 13.7. *If $\phi \leq \pi/4$ and $\langle u_1, u_2 \rangle < 0$, then*

$$v(D_4) \geq \frac{\cos \phi - \sin \phi}{\cos(\phi/2)}\omega_{n-1}.$$

Proof. Write $u = (u_1 - u_2)/\|u_1, u_2\|$ and

$$H^{**} = \{x \in E^n : \ \langle u, x \rangle = 0\}.$$

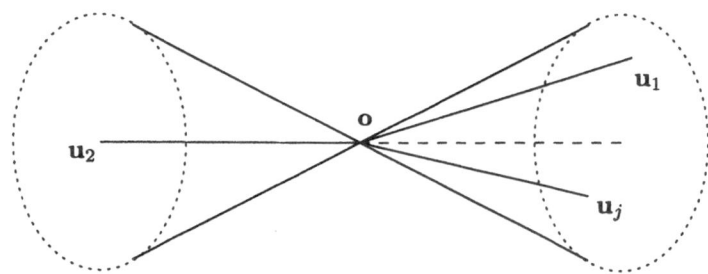

Figure 13.4

Since $\phi \leq \pi/4 < \pi/3$ it is easy to see that for $1 \leq j \leq m-1$, $\langle \mathbf{u}_1, \mathbf{u}_j \rangle \geq \cos \phi$ if and only if $\langle \mathbf{u}_2, \mathbf{u}_j \rangle \leq -\cos \phi$ (see Figure 13.4). It follows that

$$|\langle \mathbf{u}, \mathbf{u}_j \rangle| \geq \frac{|\langle \mathbf{u}_1, \mathbf{u}_j \rangle - \langle \mathbf{u}_2, \mathbf{u}_j \rangle|}{\|\mathbf{u}_1, \mathbf{u}_2\|} \geq \cos \phi,$$

which implies

$$\langle \mathbf{u}_j, \mathbf{v} \rangle \leq \sin \phi$$

for any unit vector $\mathbf{v} \in H^{**}$. Thus, for any $\lambda \in [0, 1]$ we obtain

$$\langle 2\lambda \mathbf{u}_1 + 2(1 - \lambda)\mathbf{u}_2 + \mathbf{v}, \mathbf{u}_j \rangle$$

$$\leq \left\{ \begin{array}{ll} 2\lambda(1 + \cos \phi) + 2 \cos \phi + \sin \phi, & \text{if } \langle \mathbf{u}_1, \mathbf{u}_j \rangle \geq \cos \phi, \\ -2\lambda(1 + \cos \phi) + 2 + \sin \phi, & \text{if } \langle \mathbf{u}_1, \mathbf{u}_j \rangle \leq -\cos \phi. \end{array} \right.$$

So, taking

$$h_2(\phi) = \frac{1 + \sin \phi}{2(1 + \cos \phi)}$$

and keeping our assumption in mind, for $\lambda \in [h_2(\phi), 1 - h_2(\phi)]$ we have

$$2\lambda \mathbf{u}_1 + 2(1 - \lambda)\mathbf{u}_2 \in S_n,$$

$$2\lambda \mathbf{u}_1 + 2(1 - \lambda)\mathbf{u}_2 + H^{**} \cap S_n \subset D,$$

and therefore

$$2\lambda \mathbf{u}_1 + 2(1 - \lambda)\mathbf{u}_2 + H^{**} \cap S_n \subset D^*. \qquad (13.19)$$

Let $\mathbf{w}$ be a unit vector determined by $\mathbf{w} = \alpha \mathbf{u}_1 + \beta \mathbf{u}_2$ for nonnegative α and β and satisfying $\langle \mathbf{u}, \mathbf{w} \rangle = 0$. Also, let

$$N = \{\mathbf{x} \in H^{**} \cap S_n : \langle \mathbf{w}, \mathbf{x} \rangle \geq 0\}.$$

From (13.19) and the definition of D_4 we have

$$[2\mathbf{u}_1, 2\mathbf{u}_2] + N \cap D^* \subset D_4$$

and therefore

$$v(D_4) \geq \frac{(1 - 2h_2(\phi))\|2\mathbf{u}_1, 2\mathbf{u}_2\|}{2}\omega_{n-1}$$

$$\geq \frac{\cos\phi - \sin\phi}{\cos(\phi/2)}\omega_{n-1}.$$

Lemma 13.7 is proved.   □

Lemma 13.8. *If $\phi > 0$, then for every ϵ with $0 < \epsilon < 2/\sqrt{3} - 1$ we have*

$$v(D_1) \geq \frac{h_1(1,\epsilon)^2\phi}{2(1 + h_3(1 + \epsilon, n))}\omega_{n-2},$$

where, for $1 \leq \beta < 2/\sqrt{3}$,

$$h_3(\beta, n) = \int_{1/\beta}^{1}(1 - x^2)^{(n-5)/2}dx \bigg/ \int_{\sqrt{3}/2}^{1/\beta}(1 - x^2)^{(n-5)/2}dx.$$

Proof. Using polar coordinates we have

$$v(D_1) = \frac{1}{n-2}\int_G\int_{\mathbf{x}\in H^*\cap\mathrm{bd}(S_n),\,\mathbf{y}+\mathbf{x}\in D^*}d\mathbf{x}\,d\mathbf{y}. \tag{13.20}$$

We now proceed to show that for a certain set $G^* \subset G$ with $v(G^*) > 0$, the above integral is of order ω_{n-2}. For this purpose we write

$$M_\rho = \{\mathbf{x} \in H^* \cap \mathrm{bd}(S_n):\ \rho\mathbf{x} \notin D^*\}$$

and

$$M_\rho^* = \{\mathbf{x} \in H^* \cap \mathrm{bd}(S_n):\ \rho\mathbf{x} \in D^*\},$$

and consider the inner integral at $\mathbf{y} = \mathbf{o}$. If $M_\rho = \emptyset$, then $H^* \cap \mathrm{bd}(\rho S_n)$ intersects the affine hull of certain facets F_j, say $j = 1, 2, \ldots, k$, of the Dirichlet-Voronoi cell D. Let $\mathbf{v}_j \in H^*$ be the outer unit normal of $\mathrm{aff}\{F_j\}\cap H^*$. It follows from a special case of Lemma 7.1 that

$$\mathrm{aff}\{F_j\} \cap (H^* \cap \mathrm{bd}(\rho S_n)) \subset \mathrm{int}(F_j \cap H^*)$$

for $1 \leq \rho < 2/\sqrt{3}$ and there exists an $\alpha_j \in [1, \rho]$ such that

$$\alpha_j\mathbf{v}_j \in \mathrm{int}(F_j \cap H^*).$$

Writing

$$M_j = \{\mathbf{x} \in H^* \cap \mathrm{bd}(S_n):\ \langle\mathbf{v}_j, \mathbf{x}\rangle > \alpha_j/\rho\},$$

we have

$$M_\rho = \bigcup_{j=1}^{k} M_j.$$

Defining

$$M_j^* = \left\{ \mathbf{x} \in H^* \cap \mathrm{bd}(S_n) : \sqrt{3}\alpha_j/2 \leq \langle \mathbf{v}_j, \mathbf{x} \rangle \leq \alpha_j/\rho \right\}$$

and, for $\mathbf{x} \in M_j^*$,

$$\gamma(\mathbf{x}) = \frac{\alpha_j}{\langle \mathbf{v}_j, \mathbf{x} \rangle} \geq \rho,$$

we have

$$\gamma(\mathbf{x})\mathbf{x} \in \mathrm{aff}\{F_j\} \cap H^*$$

and

$$\|\gamma(\mathbf{x})\mathbf{x}, \alpha_j\mathbf{v}_j\|^2 \leq \tfrac{4}{3} - \alpha_j^2.$$

This implies $\gamma(\mathbf{x})\mathbf{x} \in F_j \cap H^*$,

$$\mathrm{int}(M_i^*) \cap \mathrm{int}(M_j^*) = \emptyset, \quad i \neq j,$$

and

$$\bigcup_{j=1}^{k} M_j^* \subset M_\rho^*.$$

Hence, we have

$$\int_{M_\rho} d\mathbf{x} = \frac{\int_{M_\rho^*} d\mathbf{x} + \int_{M_\rho} d\mathbf{x}}{1 + \int_{M_\rho} d\mathbf{x} / \int_{M_\rho^*} d\mathbf{x}} \geq \frac{(n-2)\omega_{n-2}}{1 + \int_{M_j} d\mathbf{x} / \int_{M_j^*} d\mathbf{x}} \tag{13.21}$$

for a suitable index $j \in \{1, 2, \ldots, k\}$.

Let

$$g(x) = (1 - x^2)^{(n-5)/2},$$

and for $\xi \in [1, \rho]$ define

$$f_1(\xi) = \int_{\xi/\rho}^{1} g(x)dx$$

and

$$f_2(\xi) = \int_{\sqrt{3}\xi/2}^{\xi/\rho} g(x)dx.$$

Using polar coordinates and the monotonic property of $f_1(\xi)/f_2(\xi)$ we have

$$\frac{\int_{M_j} d\mathbf{x}}{\int_{M_j^*} d\mathbf{x}} = \frac{f_1(\alpha_j)}{f_2(\alpha_j)} \leq \frac{f_1(1)}{f_2(1)}.$$

On the other hand, it follows from (13.21) that

$$\int_{M_\rho} d\mathbf{x} \geq \frac{(n-2)\omega_{n-2}}{1 + h_3(\rho, n)}. \tag{13.22}$$

Then, applying Lemma 13.2 with $\alpha = 1$ and $\epsilon = \rho - 1$ we obtain

$$h_1(1, \epsilon)G + M_\rho^* \subseteq D^*. \tag{13.23}$$

Hence, for $0 < \epsilon < 2/\sqrt{3} - 1$, it follows from (13.20), (13.22), and (13.23) that

$$v(D_1) \geq \frac{h_1(1, \epsilon)^2 \phi}{2(1 + h_3(1 + \epsilon, n))} \omega_{n-2}.$$

This proves Lemma 13.8. $\qquad\qquad\qquad\qquad\qquad\qquad\qquad\qquad\qquad\qquad\square$

Proof of Theorem 13.2. Using Lemmas 13.4–13.8, the proof is a consequence of the fact that

$$\lim_{n \to \infty} \frac{\omega_{n-1}}{\omega_n} = \infty.$$

First we consider three cases depending on the value of ϕ and the sign of $\langle \mathbf{u}_1, \mathbf{u}_2 \rangle$.

Case 1. $0 < \phi < \pi/4$ *and* $\langle \mathbf{u}_1, \mathbf{u}_2 \rangle > 0$. By Lemmas 13.4, 13.5, and 13.6 we have

$$v(D^*) \geq v(D_1) + v(D_2) + v(D_3)$$

$$\geq \omega_{n-1} + \frac{\omega_n}{2} + \phi \left(\frac{(1 - \sin(\pi/4))^2}{2(1 - \sin^2(\pi/4))} \omega_{n-2} - \omega_{n-1} - \frac{\omega_n}{2\pi} \right)$$

$$> \omega_{n-1} + \frac{\omega_n}{2}$$

for sufficiently large n.

Case 2. $0 < \phi < \pi/4$ *and* $\langle \mathbf{u}_1, \mathbf{u}_2 \rangle < 0$. By Lemmas 13.4, 13.5, 13.6, and the relation $\cos \phi \geq 1 - \phi^2/2$ we have

$$v(D^*) \geq v(D_1) + v(D_2) + v(D_4)$$

$$\geq 2\omega_{n-1} + \phi \left(\frac{(1 - \sin(\pi/4))^2}{2(1 - \sin^2(\pi/4))} \omega_{n-2} - 2\omega_{n-1} - \frac{\pi \omega_{n-1}}{8} \right)$$

$$> 2\omega_{n-1}$$

for sufficiently large n.

Case 3. $\phi \geq \pi/4$. Choose a suitable ϵ such that the assumption of Lemma 13.8 holds. Then, by Lemma 13.8 we have

$$v(D^*) \geq v(D_1) \geq \frac{h_1(1, \epsilon)^2 \pi}{8(1 + h_3(1 + \epsilon, n))} \omega_{n-2}$$

$$> 2\omega_{n-1}$$

for sufficiently large n.

By Remark 13.2, Case 1 holds for at most two Dirichlet-Voronoi cells. Thus, when n is sufficiently large and $K_{m,n}$ is not congruent to $S_{m,n}$, we have

$$v(K_{m,n}) > \omega_n + 2(m-1)\omega_{n-1}.$$

Theorem 13.2 is proved.   □

With explicit computation, Theorem 13.2 can be restated as follows.

Theorem 13.2* (Betke, Henk, and Wills [1]). *When $n \geq 13387$, L. Fejes Tóth's sausage conjecture is true.*

Remark 13.4. *Using a more complicated method, considering a three-dimensional section instead of a two-dimensional one, Theorem 13.2* was improved by Henk [1] and by Betke and Henk [1] to $n \geq 45$ and $n \geq 42$, respectively.*

Remark 13.5. *Besides Betke, Henk, and Wills, many authors such as G. Fejes Tóth, Gritzmann, Kleinschmidt, and Pachner have made contributions towards the sausage conjecture. For example, it was proved by Gritzmann [2] that*

$$v(K_{m,n}) > \frac{\omega_n + 2(m-1)\omega_{n-1}}{2+\sqrt{2}}$$

for every n-dimensional convex body $K_{m,n}$ that contains m nonoverlapping unit spheres.

13.4. The Sausage Catastrophe

In lower dimensions, E^3 and E^4, it seems that the finite sphere packing problem is much more complicated than its high-dimensional counterpart. In 1983, Wills [1] proposed the following phenomenon.

The Sausage Catastrophe. *In E^n, $n = 3$ or 4, the sausage arrangements are optimal when the number of spheres is small; then, for a certain number k_n of spheres the extremal configurations become full-dimensional without going through intermediate-dimensional arrangements first.*

In E^3 and E^4 it was proved by Betke, Gritzmann, and Wills [1] that intermediate-dimensional arrangements are not extremal. As for the values of k_3 and k_4, it was conjectured by Wills [1] that

$$k_3 \approx 56 \quad \text{and} \quad k_4 \approx 75000.$$

In this section we demostrate several examples to support this phenomenon.

Example 13.1. Let $K_{3,n}$ be an n-dimensional convex body that contains three nonoverlapping unit spheres. Then

$$v(K_{3,n}) \geq \omega_n + 4\omega_{n-1},$$

where equality holds if and only if $K_{3,n}$ is congruent to $S_{3,n}$.

Verification. Suppose $K_{3,n}$ contains three nonoverlapping unit spheres $S_n + \mathbf{v}_1$, $S_n + \mathbf{v}_2$, and $S_n + \mathbf{v}_3$, and $v(K_{3,n})$ is minimal. Without loss of generality, we may assume that

$$\|\mathbf{v}_2, \mathbf{v}_3\| \geq \|\mathbf{v}_1, \mathbf{v}_2\| = \|\mathbf{v}_1, \mathbf{v}_3\| = 2$$

and

$$K_{3,n} = \mathrm{conv}\,\{S_n + \mathbf{v}_1, S_n + \mathbf{v}_2, S_n + \mathbf{v}_3\} = S_n + T,$$

where T is the triangle with vertices $\mathbf{v}_1$, $\mathbf{v}_2$, and $\mathbf{v}_3$.

Clearly, since S_i is the largest section of S_{i+1}, we have

$$\omega_{i+1} < 2\omega_i.$$

Then, writing

$$\theta = \min\,\{\angle \mathbf{v}_1\mathbf{v}_2\mathbf{v}_3, \angle \mathbf{v}_1\mathbf{v}_3\mathbf{v}_2\},$$

we have $0 \leq \theta \leq \pi/3$ and

$$
\begin{aligned}
v(K_{3,n}) &= s(T)\omega_{n-2} + \tfrac{1}{2}(4 + \|\mathbf{v}_2, \mathbf{v}_3\|)\omega_{n-1} + \omega_n \\
&= 4\sin\theta\cos\theta\,\omega_{n-2} + 2(1 + \cos\theta)\omega_{n-1} + \omega_n \\
&> 2(1 + \cos\theta + \sin\theta\cos\theta)\omega_{n-1} + \omega_n.
\end{aligned}
$$

It is easy to see that

$$f(x) = 1 + \cos x + \sin x \cos x$$

attains its minimum, 2, only at $x = 0$, for $0 \leq x \leq \pi/3$. Thus, we have

$$v(K_{3,n}) \geq \omega_n + 4\omega_{n-1},$$

and equality can be attained if and only if $K_{3,n}$ is congruent to $S_{3,n}$. The example is verified. □

Example 13.1 shows that the sausage arrangements are optimal for three spheres in E^n, $n \geq 3$. In E^3, this result was extended to four spheres by Böröczky Jr. [1]. On the other hand, Gandini and Wills [1] showed that in E^3 for $m = 56$, 59–62, and all $m \geq 65$ some full-dimensional packings of spheres are best possible and conjectured that in all remaining cases the sausage packing is optimal. The following example considers the case $m = 56$.

Example 13.2 (Gandini and Wills [1]). There is a three-dimensional convex body $K_{56,3}$ such that

$$v(K_{56,3}) < v(S_{56,3}) = \omega_3 + 110\omega_2.$$

Verification. Write $\mathbf{u}_1 = (\sqrt{2}, \sqrt{2}, 0)$, $\mathbf{u}_2 = (\sqrt{2}, 0, \sqrt{2})$, $\mathbf{u}_3 = (0, \sqrt{2}, \sqrt{2})$, and define

$$\Lambda = \left\{ \sum_{i=1}^{3} z_i \mathbf{u}_i \ : \ z_i \in Z \right\}$$

and

$$T_k = \text{conv} \left\{ \mathbf{o}, k\mathbf{u}_1, k\mathbf{u}_2, k\mathbf{u}_3 \right\}.$$

Then, $S_3 + \Lambda$ is a lattice packing and

$$\text{card} \left\{ T_k \cap \Lambda \right\} = \binom{k+3}{3}. \tag{13.24}$$

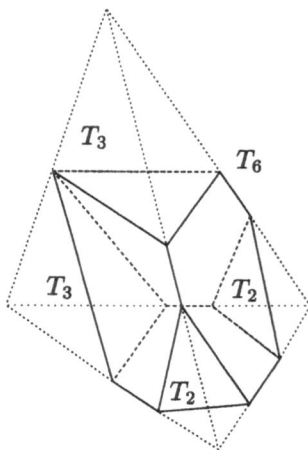

Figure 13.5

Let K be the convex polytope obtained by cutting off two translates of T_2 and two translates of T_3 from T_6 (see Figure 13.5). Then, by (13.24) and routine computation we obtain

$$\text{card}\{K \cap \Lambda\} = 56.$$

On the other hand, by routine computation we have

$$v(S_3 + K) < \omega_3 + 110\omega_2 = v(S_{56,3}).$$

Thus, taking

$$K_{56,3} = S_3 + K,$$

Example 13.2 follows.   □

In E^4, applying a similar method to a 24-cell with side length 34, we obtain the following result.

Example 13.3 (Gandini and Zucco [1]). There is a four-dimensional convex body $K_{375769,4}$ such that

$$v(K_{375769,4}) < v(S_{375769,4}) = \omega_4 + 2 \times 375768\omega_3.$$

Based on this result, Gandini and Zucco [1] gave the following modified version of Wills' conjecture:

$$k_4 \approx 375769.$$

Bibliography

Afflerbach, L. [1]. Minkowskische Reduktionsbedingungen für positive definite quadratische Formen in 5 Variablen, *Monatsh. Math.* **94** (1982), 1–8.

Aigner, M. and Ziegler, G. [1]. *Proofs from the Book*, Springer-Verlag, Berlin, 1998.

Alon, N. [1]. Packings with large minimum kissing numbers, *Discrete Math.* **175** (1997), 249–251.

Babai, L. [1]. On Lovász' lattice reduction and the nearest lattice point problem, *Combinatorica* **6** (1986), 1–13.

Ball, K. [1]. Ellipsoids of maximal volume in convex bodies, *Geom. Dedicata* **41** (1992), 241–250. [2]. A lower bound for the optimal density of lattice packings, *Internat. Math. Res. Notices* **10** (1992), 217–221.

Bambah, R.P. [1]. On lattice coverings by spheres, *Proc. Nat. Inst. Sci. India* **20** (1954), 25–52. [2]. Lattice coverings with four-dimensional spheres, *Proc. Camb. Phil. Soc.* **50** (1954), 203–208.

Bambah, R.P. and Davenport, H. [1]. The covering of n-dimensional space by spheres, *J. London Math. Soc.* **27** (1952), 224–229.

Bambah, R.P. and Roth, K.F. [1]. Lattice coverings, *J. Indian Math. Soc.* **16** (1952), 7–12.

Bambah, R.P. and Woods, A.C. [1]. The thinnest double lattice covering of three-spheres, *Acta Arith.* **18** (1971), 321–336.

Bannai, E. and Sloane, N.J.A. [1]. Uniqueness of certain spherical codes, *Canad. J. Math.* **33** (1981), 437–449.

Baranovskii, E.P. [1]. On packing n-dimensional Euclidean spaces by equal spheres, *Izv. Vysš. Učebn. Zved. Mat.* **39** (1964), 14–24. [2]. Packings, coverings, partitionings and certain other distributions in spaces of constant curvature, *Progress Math.* **9** (1969), 209–253.

Bárány, I. and Leader, I. [1]. Private communication.

Barlow, W. [1]. Probable nature of the internal symmetry of crystals, *Nature* **29** (1883), 186–188.

Barnes, E.S. [1]. Note on extreme forms, *Canad. J. Math.* **7** (1955), 145–149. [2]. The covering of space by spheres, *Canad. J. Math.* **8** (1956), 293–304. [3]. The perfect and extreme senary forms, *Canad. J. Math.* **9** (1957), 235–242. [4]. The complete enumeration of extreme senary forms, *Phil. Transact. Royal Soc. London* (A) **249** (1957), 461–506. [5]. Criteria for extreme forms, *J. Austral. Math. Soc.* **1** (1959), 17–20. [6]. The construction of perfect and extreme forms I, *Acta Arith.* **5** (1959), 57–79. [7]. The construction of perfect and extreme forms II, *Acta Arith.* **5** (1959), 205–222.

Barnes, E.S. and Dickson, T.J. [1]. Extreme coverings of n-space by spheres, *J. Austral. Math. Soc.* **7** (1967), 115–127; Corrigendum **8**, 638–640. [2]. The extreme covering of 4-space by spheres, *J. Austral. Math. Soc.* **7** (1967), 490–496.

Barnes, E.S. and Sloane, N.J.A. [1]. New lattice packings of spheres, *Canad. J. Math.* **35** (1983), 117–130.

Barnes, E.S. and Trenerry, D.W. [1]. A class of extremal lattice coverings of n-space by spheres, *J. Austral. Math. Soc.* **14** (1972), 247–256.

Bassalygo, L.A. and Zinoviev, V.A. [1]. Some simple consequences of coding theory for combinatorial problems of packings and coverings, *Problems Inform. Transmission* **34** (1983), 629–631.

Bateman, P.T. [1]. The Minkowski-Hlawka theorem in the geometry of numbers, *Arch. Math.* **13** (1962), 357–362.

Bender, C. [1]. Bestimmung der grössten Anzahl gleich grosser Kugeln, welche sich auf eine Kugel von demselben Radius, wie die übrigen, auflegen lassen, *Grunert Arch. Math. Phys.* **56** (1874), 302–306.

Bergé, A.M. and Martinet, J. [1]. Sur un problème de dualité lié aux sphères en géométrie des nombres, *J. Number Theory* **32** (1989), 14–42.

Betke, U. and Gritzmann, P. [1]. Über L. Fejes Tóth's Wurstvermutung in kleinen Dimensionen, *Acta Math. Acad. Sci. Hungar.* **43** (1984), 299–307.

Betke, U., Gritzmann, P., and Wills, J.M. [1]. Slices of L. Fejes Tóth's sausage conjecture, *Mathematika* **29** (1982), 194–201.

Betke, U. and Henk, M. [1]. Finite packings of spheres, *Discrete & Comput. Geom.* **19** (1998), 197–227. [2]. Densest lattice packings of 3-polytopes, preprint.

Betke, U., Henk, M., and Wills, J.M. [1]. Finite and infinite packings, *J. reine angew. Math.* **453** (1994), 165–191. [2]. Sausages are good packings, *Discrete & Comput. Geom.* **13** (1995), 297–311.

Bezdek, A. and Kuperberg, W. [1]. Packing Euclidean space with congruent cylinders and with congruent ellipsoids, *Applied Geometry and Discrete Mathematics*, Amer. Math. Soc., Providence 1991, 71–80.

Bezdek, A., Kuperberg, W., and Makai Jr., E. [1]. Maximum density space packing with parallel strings of balls, *Discrete & Comput. Geom.* **6** (1991), 277–283.

Bezdek, K. [1]. Isoperimetric inequalities and the dodecahedral conjecture, *Internat. J. Math.* **8** (1997), 759–780.

Bieberbach, L. and Schur, I. [1]. Über die Minkowskische Reduktionstheorie der positiven quadratischen Formen, *S.-B. Preuss. Akad. Wiss.* (1928), 510–535.

Blachman, N.M. and Few, L. [1]. Multiple packing of spherical caps, *Mathematika* **10** (1963), 84–88.

Bleicher, M.N. [1]. Lattice covering of n-space by spheres, *Canad. J. Math.* **14** (1962), 632–650.

Blichfeldt, H.F. [1]. The minimum value of quadratic forms and the closest packing of spheres, *Math. Ann.* **101** (1929), 605–608. [2]. The minimum values of positive quadratic forms in six, seven and eight variables, *Math. Z.* **39** (1934), 1–15. [3]. A new upper bound to the minimum value of the sum of linear homogeneous forms, *Monatsh. Math.* **43** (1936), 410–414.

Blundon, W.J. [1]. Multiple covering of the plane by circles, *Mathematika* **4** (1957), 29–31. [2]. Multiple packing of circles in the plane, *J. London Math. Soc.* **38** (1963), 176–182. [3]. Some lower bounds for density of multiple packing, *Canad. Math. Bull.* **7** (1964), 565–572.

Böhm, J. and Hertel, E. [1]. *Polyedergeometrie in n-dimensionalen Räumen konstanter Krümmung*, VEB Deutsch. Verlag Wiss., Berlin, 1980.

Bolle, U. [1]. Dichteabschätzungen für mehrfache gitterförmige Kugelanordnungen in R^n, *Studia Sci. Math. Hungar.* **17** (1982), 429–444. [2]. Über die Dichte mehrfacher gitterförmiger Kreisanordnungen in der Ebene, *Studia Sci. Math. Hungar.* **19** (1983), 275–284. [3]. On the density of multiple packings and coverings of convex discs, *Studia Sci. Math. Hungar.* **24** (1989), 119–126.

Boltjanski, V. and Gohberg, I. [1]. *Results and Problems in Combinatorial Geometry*, Cambridge University Press, Cambridge, 1985.

Böröczky, K. [1]. Packing of spheres in space of constant curvature, *Acta Math. Acad. Sci. Hungar.* **32** (1978), 243–261. [2]. Closest packing and loosest covering of the space with balls, *Studia Sci. Math. Hungar.* **21** (1986), 79–89.

Böröczky, K. and Soltan, V. [1]. Translational and homothetic clouds for a convex body, *Studia Sci. Math. Hungar.* **32** (1996), 93–102.

Böröczky Jr., K. [1]. About four-ball packings, *Mathematika* **40** (1993), 226–232. [2]. Mean projections and finite packings of convex bodies, *Monatsh. Math.* **118** (1994), 41–54.

Böröczky Jr., K. and Henk, M. [1]. Radii and the sausage conjecture, *Canad. Math. Bull.* **38** (1995), 156–166. [2]. Random projections of regular polytopes, *Arch. Math.*, in press.

Böröczky Jr., K. and Schnell, U. [1]. Asymptotic shape of finite packings, *Canad. J. Math.*, in press.

Böröczky Jr., K. and Wills, J.M. [1]. Finite sphere packings and critical radii, *Beiträge Alg. & Geom.* **38** (1997), 193–211.

Bos, A., Conway, J.H., and Sloane, N.J.A. [1]. Further lattice packings in high dimensions, *Mathematika* **29** (1982), 171–180.

Bourgain, J. [1]. On lattice packing of convex symmetric sets in R^n, *Lecture Notes Math.*, Springer-Verlag, Berlin, **1267** (1986), 5–12.

Boyvalenkov, P. [1]. Extremal polynomials for obtaining bounds for spherical codes and designs, *Discrete & Comput. Geom.* **14** (1995), 277–286.

Butler, G.J. [1]. Simultaneous packing and covering in Euclidean space, *Proc. London Math. Soc.* **25** (1972), 721–735.

Cassels, J.W.S. [1]. A short proof of the Minkowski-Hlawka theorem, *Proc. Cambridge Phil. Soc.* **49** (1953), 165–166. [2]. *An Introduction to the Geometry of Numbers*, Springer-Verlag, Berlin, 1959.

Chabauty, C. [1]. Empilement de sphères égales dans R^n et valeur asymptotique de la constante γ_n d'Hermite, *C. R. Acad. Sci. Paris* **235** (1952), 529–532.

Charve, H.F. [1]. De la réduction des formes quadratiques quaternaires positives, *Ann. Sci. École Norm. Sup.* **11** (1882), 119–142.

Churchhouse, R.F. [1]. An extension of the Minkowski-Hlawka theorem, *Proc. Cambridge Phil. Soc.* **50** (1954), 220–224.

Cleaver, F.L. [1]. On a theorem of Voronoi, *Trans. Amer. Math. Soc.* **120** (1965), 390–400. [2]. On coverings of four-space by spheres, *Trans. Amer. Math. Soc.* **120** (1965), 402–416.

Cohen, A.M. [1]. Numerical determination of lattice constants, *J. London Math. Soc.* **37** (1962), 185–188.

Cohn, M.J. [1]. Multiple lattice covering of space, *Proc. London Math. Soc.* **32** (1972), 117–132.

Connelly, R. [1]. Rigid circle and sphere packings I, finite packings, *Structural Topology* **14** (1988), 43–60.

Conway, J.H. [1]. A characterization of Leech's lattice, *Invent. Math.* **7** (1969), 137–142. [2]. Sphere packings, lattices, codes, and greed, *Proc. Internat. Congr. Math. Zurich*, Birkhäuser, Basel, 1995, 45–55.

Conway, J.H., Duff, T.D.S., Hardin, R.H., and Sloane, N.J.A. [1]. Minimal energy clusters of hard spheres, *Discrete & Comput. Geom.* **14** (1995), 237–259.

Conway, J.H., Parker, R.A., and Sloane, N.J.A. [1]. The covering radius of the Leech lattice, *Proc. Royal Soc. London* (A) **380** (1982), 261–290.

Conway, J.H. and Sloane, N.J.A. [1]. *Sphere Packings, Lattices and Groups*, Springer-Verlag, New York, 1998. [2]. What are all the best sphere packings in low dimensions? *Discrete & Comput. Geom.* **13** (1995), 383–403.

van der Corput, J.G. and Schaake, G. [1]. Anwendung einer Blichfeldtschen Beweismethode in der Geometrie der Zahlen, *Acta Math.* **2** (1937), 152–160.

Coxeter, H.S.M. [1]. Extreme forms, *Canad. J. Math.* **3** (1951), 391–441. [2]. Arrangements of equal spheres in non-Euclidean spaces, *Acta Math. Acad. Sci. Hungar.* **5** (1954), 263–274. [3]. Close-packing and froth, *Illinois J. Math.* **2** (1958), 746–758. [4]. An upper bound for the number of equal nonoverlapping spheres that can touch another of the same size, *Proc. Symp. Pure Math.* **7** (1963), 53–71. [5]. *Regular Polytopes*, Dover Publications Inc., New York, 1973.

Coxeter, H.S.M., Few, L., and Rogers, C.A. [1]. Covering space with equal spheres, *Mathematika* **6** (1959), 147–157.

Croft, H.T., Falconer, H.J., and Guy, R.K. [1]. *Unsolved Problems in Geometry*, Springer-Verlag, New York, 1991.

Csóka, G. [1]. The number of congruent spheres that hide a given sphere of three-dimensional space is not less than 30, *Studia Sci. Math. Hungar.* **12** (1977), 323–334.

Dalla, L., Larman, D.G., Mani-Levitska, P., and Zong, C. [1]. The blocking numbers of convex bodies, *Discrete & Comput. Geom.*, in press.

Danzer, L. [1]. Drei Beispiele zu Lagerungsproblemen, *Arch. Math.* **11** (1960), 159–165.

Dauenhauer, M.H. and Zassenhaus, H.J. [1]. Local optimality of the critical lattice sphere packing of regular tetrahedra, *Discrete Math.* **64** (1987), 129–146.

Davenport, H. [1]. Sur un système de sphères qui recouvrent l'espace à *n* dimensions, *C. R. Acad. Sci. Paris* **233** (1951), 571–573. [2]. The covering of space by spheres, *Rend. Circ. Mat. Palermo* **1** (1952), 92–107. [3]. Problems of packing and covering, *Univ. e Politec. Torino Rend. Sem. Mat.* **24** (1964/65), 41–48.

Davenport, H. and Hajös, G. [1]. Problem 35, *Mat. Lapok* **2** (1951), 68.

Davenport, H. and Rogers, C.A. [1]. Hlawka's theorem in the geometry of numbers, *Duke Math. J.* **14** (1947), 367–375.

Davenport, H. and Watson, G.L. [1]. The minimal points of a positive definite quadratic form, *Mathematika* **1** (1954), 14–17.

Delone, B.N. [1]. Sur la sphère vide, *Proc. Internat. Congr. Math. Toronto* (1924), 695–700. [2]. About the closest lattice packing of balls in the space of 3 and 4 dimensions, *Trudy Fiz-Mat. Inst. Otdel. Math.* **4** (1933), 63–69. [3]. The geometry of positive quadratic forms I, *Uspehi Mat. Nauk* **3** (1937), 16–62. [4]. The geometry of positive quadratic forms II, *Uspehi Mat. Nauk* **4** (1938), 102–167.

Delone, B.N. and Ryškov, S.S. [1]. Solution of the problem on the least dense lattice covering of a 4-dimensional space by equal spheres, *Soviet Math. Dokl.* **4** (1963), 1333–1334.

Delsarte, P. [1]. Bounds for unrestricted codes by linear programming, *Philips Res. Rep.* **27** (1972), 272–289. [2]. An algebraic approach to the association schemes of coding theory, *Philips Res. Rep. Suppl.* **27** (1973), 1–97.

Delsarte, P., Goethals, J.M., and Seidel, J.J. [1]. Spherical codes and designs, *Geom. Dedicata* **6** (1977), 363–388.

Dickson, T.J. [1]. The extreme coverings of 4-space by spheres, *J. Austral. Math. Soc.* **7** (1967), 490–496. [2]. A sufficient condition for extreme covering of *n*-space by spheres, *J. Austral. Math. Soc.* **8** (1968), 56–62. [3]. On Voronoi reduction of positive definite quadratic forms, *J. Number Theory* **4** (1972), 330–341.

Dirichlet, G.L. [1]. Über die Reduction der positiven quadratischen Formen mit drei unbestimmten ganzen Zahlen, *J. reine angew. Math.* **40** (1850), 209–227.

Dumir, V.C. [1]. Lattice double coverings by spheres, *Proc. Nat. Inst. Sci. India* **33** (1967), 259–263.

Elkies, N.D., Odlyzko, A.M., and Rush, J.A. [1]. On the packing density of superballs and other bodies, *Invent. Math.* **105** (1991), 613–639.

Erdahl, R.M. and Ryškov, S.S. [1]. The empty sphere, *Canad. J. Math.* **39** (1987), 794–824; **40** (1988), 1058–1073.

Erdös, P., Gruber, P.M., and Hammer, J. [1]. *Lattice Points*, Longman Scientific & Technical, Essex, 1989.

Erdös, P. and Rogers, C.A. [1]. The covering of n-dimensional space by spheres, *J. London Math. Soc.* **28** (1953), 287–293.

Fejes Tóth, G. [1]. Multiple packing and covering of the plane with circles, *Acta Math. Acad. Sci. Hungar.* **27** (1976), 135–140. [2]. Multiple packing and covering of spheres, *Acta Math. Acad. Sci. Hungar.* **34** (1979), 165–176. [3]. Densest packings of typical convex sets are not lattice-like, *Discrete & Comput. Geom.* **14** (1995), 1–8.

Fejes Tóth, G. and Florian, A. [1]. Mehrfache gitterförmige Kreis- und Kugelanordnungen, *Monatsh. Math.* **79** (1975), 13–20.

Fejes Tóth, G., Gritzmann, P., and Wills, J.M. [1]. Sausage-skin problems for finite coverings, *Mathematika* **31** (1984), 118–137. [2]. Finite sphere packing and sphere covering, *Discrete & Comput. Geom.* **4** (1989), 19–40. [3]. On finite multiple packings, *Arch. Math.* **54** (1990), 407–411.

Fejes Tóth, G. and Kuperberg, W. [1]. Packing and covering with convex sets, *Handbook of Convex Geometry* (P.M. Gruber and J.M. Wills eds.), North-Holland, Amsterdam, 1993, 799–860. [2]. Blichfeldt's density bound revisited, *Math. Ann.* **295** (1993), 721–727.

Fejes Tóth, L. [1]. Über einen geometrischen Satz, *Math. Z.* **46** (1940), 83–85. [2]. Über die dichteste Kugellagerung, *Math. Z.* **48** (1943), 676–684. [3]. Über eine Abschätzung des kürzesten Abstandes zweier Punkte eines auf eine Kugelfläche liegenden Punktsystems, *Jber. Deutsch. Math. Verein.* **53** (1943), 66–68. [4]. Über dichteste Kreislagerung und dünnste Kreisüberdeckung, *Comment. Math. Helv.* **23** (1949), 342–349. [5]. Some packing and covering theorems, *Acta Math. Acad. Sci. Hungar.* **12** (1950), 62–67. [6]. On close packing of spheres in spaces of constant curvature, *Publ. Math. Debrecen* **3** (1953), 158–167. [7]. Verdeckung einer Kugel durch Kugeln, *Publ. Math. Debrecen* **6** (1959), 234–240. [8]. Kugelunterdeckungen und Kugelüberdeckungen in Räumen konstanter Krümmung, *Arch. Math.* **10** (1959), 307–313. [9]. *Lagerungen in der Ebene, auf der Kugel und im Raum*, Springer-Verlag, Berlin, 1971. [10]. Research problem 13, *Period. Math. Hungar.* **6** (1975), 197–199. [11]. Close packing and loose covering with balls, *Publ. Math. Debrecen* **23** (1976), 323–326. [12]. On the densest packing of convex discs, *Mathematika* **30** (1983), 1–3.

Ferguson, S.P. [1]. Spere packings V, preprint.

Ferguson, S.P. and Hales, T.C. [1]. A formulation of the Kepler conjecture, preprint.

Few, L. [1]. The double packing of spheres, *J. London Math. Soc.* **28** (1953), 297–304. [2]. Covering space by spheres, *Mathematika* **3** (1956), 136–139. [3]. Multiple packing of spheres, *J. London Math. Soc.* **39** (1964), 51–54. [4]. Multiple packings of spheres: a survey, *Proc. Colloq. Convexity*, Copenhagen, 1965, 86–93. [5]. Double covering with spheres, *Mathematika* **14** (1967), 207–214. [6]. Double packing of spheres: a new upper bound, *Mathematika* **15** (1968), 88–92.

Few, L. and Kanagasabapathy, P. [1]. The double packing of spheres, *J. London Math. Soc.* **44** (1969), 141–146.

Florian, A. [1]. Ausfüllung der Ebene durch Kreise, *Rend. Circ. Mat. Palermo* **9** (2) (1960), 1–13. [2]. Zur Geometrie der Kreislagerungen, *Acta Math. Acad. Sci. Hungar.* **18** (1967), 341–358.

Gamečkii, A.F. [1]. On the theory of covering Euclidean n-space by equal spheres, *Soviet Math. Dokl.* **3** (1962), 1410–1414.

Gandini, P.M. and Wills, J.M. [1]. On finite sphere packings, *Math. Pannon* **3** (1992), 19–29.

Gandini, P.M. and Zucco, A. [1]. On the sausage catastrophe in 4-space, *Mathematika* **39** (1992), 274–278.

Gauss, C.F. [1]. Untersuchungen über die Eigenschaften der positiven ternären quadratischen Formen von Ludwig August Seeber, *J. reine angew. Math.* **20** (1840), 312–320.

Golay, M.J.E. [1]. Binary coding, *IEEE Trans. Inform. Theory* **4** (1954), 23–28.

Goldberg, M. [1]. On the densest packing of equal spheres in a cube, *Math. Mag.* **44** (1971), 199–208.

Gritzmann, P. [1]. *Finite Packungen und Überdeckungen*, Habilitationsschrift, Universität Siegen, 1984. [2]. Finite packing of equal balls, *J. London Math. Soc.* **33** (1986), 543–553.

Gritzmann, P. and Wills, J.M. [1]. Finite packing and covering, *Studia Sci. Math. Hungar.* **21** (1986), 149–162. [2]. Finite packing and covering, *Handbook of Convex Geometry* (P.M. Gruber and J.M. Wills eds.), North-Holland, Amsterdam, 1993, 861–897.

Groemer, H. [1]. Über die Einlagerung von Kreisen in einem konvexen Bereich, *Math. Z.* **73** (1960), 285–294. [2]. Abschätzungen für die Anzahl der konvexen Körper, die einen konvexen Körper berühren, *Monatsh. Math.* **65** (1961), 74–81. [3]. Über die dichteste gitterförmige Lagerung kongruenter Tetraeder, *Monatsh. Math.* **66** (1962), 12–15. [4]. Multiple packings and coverings, *Studia Sci. Math. Hungar.* **21** (1986), 189–200. [5]. *Geometric Applications of Fourier Series and Spherical Harmonics*, Cambridge University Press, Cambridge, 1996.

Grötschel, M., Lovász, L., and Schrijver, A. [1]. *Geometric Algorithms and Combinatorial Optimization*, Springer-Verlag, Berlin, 1988.

Gruber, P.M. [1]. Typical convex bodies have surprising few neighbors in densest lattice packings, *Studia Sci. Math. Hungar.* **21** (1986), 163–173. [2]. Geometry of numbers, *Handbook of Convex Geometry* (P.M. Gruber and J.M. Wills eds.), North-Holland, Amsterdam, 1993, 739–763.

Gruber, P.M. and Lekkerkerker, C.G. [1]. *Geometry of Numbers*, North-Holland, Amsterdam, 1987.

Grünbaum, B. [1]. On a conjecture of H. Hadwiger, *Pacific J. Math.* **11** (1961), 215–219.

Günther, S. [1]. Ein stereometrisches Problem, *Grunert Arch. Math. Phys.* **57** (1875), 209–215.

Hadwiger, H. [1]. Über Treffanzahlen bei translationsgleichen Eikörpern, *Arch. Math.* **8** (1957), 212–213. [2]. *Vorlesungen über Inhalt, Oberfläche und Isoperimetrie*, Springer-Verlag, Berlin, 1957. [3]. Überdeckung des Raumes durch translationsgleiche Punktmengen und Nachbarzahlen, *Monatsh. Math.* **73** (1969), 213–217.

Halberg, C., Levin, E., and Straus, E. [1]. On contiguous congruent sets in Euclidean space, *Proc. Amer. Math. Soc.* **10** (1959), 335–344.

Hales, T.C. [1]. The sphere packing problem, *J. Comput. Appl. Math.* **44** (1992), 41–76. [2]. Remarks on the density of sphere packings in three dimensions, *Combinatorica* **13** (1993), 181–197. [3]. The status of the Kepler conjecture, *Math. Intelligencer* **16** (3) (1994), 47–58. [4]. Sphere packings I, *Discrete & Comput. Geom.* **17** (1997), 1–51. [5]. Sphere packings II, *Discrete & Comput. Geom.* **18** (1997), 135–149. [6]. Sphere packings III, preprint. [7]. Sphere packings IV, preprint.

Hardy, G.H., Littlewood, J.E., and Pólya, G. [1]. *Inequalities*, Cambridge University Press, Cambridge, 1934.

Henk, M. [1]. *Finite and Infinite Packings*, Habilitationsschrift, Universität Siegen, 1995. [2]. Private communication.

Heppes, A. [1]. Mehrfache gitterförmige Kreislagerungen in der Ebene, *Acta Math. Akad. Sci. Hungar.* **10** (1959), 141–148. [2]. Ein Satz über gitterförmige Kugelpackungen, *Ann. Univ. Sci. Budapest Sect. Math.* **3-4** (1960/61), 89–90. [3]. On the number of spheres which can hide a given sphere, *Canad. J. Math.* **19** (1967), 413–418.

Hermite, Ch. [1]. Extraits de lettre de M. Ch. Hermite à M. Jacobi sur differents objets de la théorie des nombres, *J. reine angew. Math.* **40** (1850), 261–278.

Hilbert, D. [1]. Mathematische Probleme, *Arch. Math. Phys.* **1** (1901), 44–63.

Hlawka, E. [1]. Zur Geometrie der Zahlen, *Math. Z.* **49** (1943), 285–312. [2]. Über potenzsummen von Linearformen, *Sitzungsber. Österr. Akad. Wiss., Math.-Naturwiss. KL* **154** (1945), 50–58. [3]. Über potenzsummen von Linearformen II, *Sitzungsber. Österr. Akad. Wiss., Math.-Naturwiss. KL* **156** (1947), 247–254. [4]. Ausfüllung und Überdeckung konvexer Körper durch konvexe Körper, *Monatsh. Math. Phys.* **53** (1949), 81–131. [5]. *Edmund Hlawka Selecta* (P.M. Gruber and W.M. Schmidt eds), Springer-Verlag, Berlin, Heidelberg, 1990.

Hofreiter, N. [1]. Über Extremformen, *Monatsh. Math. Phys.* **40** (1933), 129–152.

Hoppe, R. [1]. Bemerkungen der Redaktion, *Grunert Arch. Math. Phys.* **56** (1874), 307–312.

Hortobágyi, I. [1]. Durchleuchtung gitterförmiger Kugelpackungen mit Lichtbündeln, *Studia Sci. Math. Hungar.* **6** (1971), 147–150.

Horváth, J. [1]. Über die Durchsichtigkeit gitterförmiger Kugelpackungen, *Studia Sci. Math. Hungar.* **5** (1970), 421–426. [2]. On close lattice packing of unit spheres in the space E^n, *Proc. Steklov Math. Inst.* **152** (1982), 237–254. [3]. Eine Bemerkung zur Durchleuchtung von gitterförmigen Kugelpackungen, *Proc. Internat. Conf. Geom.*, Thessalohiki, 1996, 187–191.

Hoylman, D.J. [1]. The densest lattice packing of tetrahedra, *Bull. Amer. Math. Soc.* **76** (1970), 135–137.

Hsiang, W.Y. [1]. A simple proof of a theorem of Thue on the maximal density of circle packings in E^2, *Enseign. Math.* **38** (1992), 125–131. [2]. On the sphere packing problem and the proof of Kepler's conjecture, *Internat. J. Math.* **4** (1993), 739–831. [3]. A rejoinder to Hales' article, *Math. Intelligencer* **17** (1) (1995), 35–42.

Hua, L.K. [1]. A remark on a result due to Blichtfeldt, *Bull. Amer. Math. Soc.* **51** (1945), 537–539.

Ignatev, N.K. [1]. On a practical method for finding dense packings of n-dimensional spheres, *Siber. Math. J.* **5** (1965), 815–819. [2]. On finding dense packings of n-dimensional spheres, *Siber. Math. J.* **7** (1967), 653–657.

John, F. [1]. Extremum problems with inequalities as subsidiary conditions, *Courant Ann. Volume*, Interscience, New York, 1948, 187–204.

Kabatjanski, G.A. and Levenštein, V.I. [1]. Bounds for packings on a sphere and in space, *Problems Inform. Transmission* **14** (1978), 1–17.

Kepler, J. [1]. *On the six-cornered snowflake,* Oxford University Press, Oxford, 1966.

Kershner, R. [1]. The number of circles covering a set, *Amer. J. Math.* **61** (1939), 665–671.

Kleinschmidt, P., Pachner, U., and Wills, J.M. [1]. On L. Fejes Tóth's sausage conjecture, *Israel J. Math.* **47** (1984), 216–226.

Kneser, M. [1]. Lineare Relationen zwischen Darstellungensanzahlen quadratischer Formen, *Math. Ann.* **168** (1967), 31–39.

Korkin, A.N. and Zolotarev, E.I. [1]. Sur les formes quadratiques positives quaternaires, *Math. Ann.* **5** (1872), 581–583. [2]. Sur les formes quadratiques, *Math. Ann.* **6** (1873), 366–389. [3]. Sur les formes quadratiques positives, *Math. Ann.* **11** (1877), 242–292.

Lagarias, J.C., Lenstra Jr., H.W., and Schnorr, C.P. [1]. Korkine-Zolotarev bases and successive minima of a lattice and its reciprocal lattice, *Combinatorica* **10** (1990), 333–348.

Lagrange, J.L. [1]. Recherches d'arithmétique, *Nouveaux Mémoires de L'Académie royal des Sciences et Belles-Lettres de Berlin* (1773), 265–312.

Larman, D.G. and Zong, C. [1]. The kissing numbers of some special convex bodies, *Discrete & Comput. Geom.* **21** (1999), 233–242.

Leech, J. [1]. The problem of the thirteen spheres, *Math. Gazette* **40** (1956), 22–23. [2]. Some sphere packings in higher space, *Canad. J. Math.* **16** (1964), 657–682. [3]. Note on sphere packings, *Canad. J. Math.* **19** (1967), 251–267. [4]. Five dimensional non-lattice sphere packings, *Canad. Math. Bull.* **10** (1967), 387–393. [5]. Six and seven dimensional non-lattice sphere packings, *Canad. Math. Bull.* **12** (1969), 151–155.

Leech, J. and Sloane, N.J.A. [1]. New sphere packings in dimensions 9–15, *Bull. Amer. Math. Soc.* **76** (1970), 1006–1010. [2]. Sphere packings and error-correcting codes, *Canad. J. Math.* **23** (1971), 718–745.

Leichtweiß, K. [1]. Über die affine Exzentrizität konvexer Körper, *Arch. Math.* **10** (1959), 187–199. [2]. *Konvexe Mengen,* Springer-Verlag, Berlin, 1980.

Lekkerkerker, C.G. [1]. On the Minkowski-Hlawka theorem, *Proc. Kon. Ned. Akad. Wet.* **59** (1956), 426–434.

Lenstra, A.K., Lenstra Jr, H.W., and Lovász, L. [1]. Factoring polynomials with rational coefficients, *Math. Ann.* **261** (1982), 515–534.

Leppmeier, M. [1]. *Kugelpackungen von Kepler bis heute,* Vieweg, Braunschweig, 1997.

Levenštein, V.I. [1]. The maximal density of filling an n-dimensional Euclidean space with equal balls, *Math. Notes* **18** (1975), 765–771. [2]. On bounds for packings in n-dimensional Euclidean space, *Soviet Math. Dokl.* **20** (1979), 417–421. [3]. Bounds for packings of metric spaces and some of their applications, *Prob. Kibern.* **40** (1983), 43–110.

Lindsey II, J.H. [1]. Sphere packing in R^3, *Mathematika* **33** (1986), 137–147.

Linhart, J. [1]. Eine Methode zur Berechnung der Dichte einer dichtesten gitterförmigen k-fachen Kreispackung, *Arbeitsber. Math. Inst. Univ. Salzburg*, 1983.

Litsyn, S.N. and Tsfasman, M.A. [1]. Algebraic-geometric and number-theoretic packings of spheres, *Uspekhi Mat. Nauk.* **40** (1985), 185–186. [2]. Constructive high-dimensional sphere packings, *Duke Math. J.* **54** (1987), 147–161.

Litsyn, S.N. and Vardy, A. [1]. The uniqueness of the best code, *IEEE Trans. Inform. Theory* **40** (1994), 1693–1698.

Lloyd, S.P. [1]. Hamming association schemes and codes on spheres, *SIAM J. Math. Anal.* **11** (1980), 488–505.

Macbeath, A.M. and Rogers, C.A. [1]. A modified form of Siegel's mean value theorem, *Proc. Cambridge Phil. Soc.* **51** (1955), 565–576. [2]. A modified form of Siegel's mean value theorem II, *Proc. Cambridge Phil. Soc.* **54** (1958), 322–326. [3]. Siegel's mean value theorem in the geometry of numbers, *Proc. Cambridge Phil. Soc.* **54** (1958), 139–151.

MacWilliams, F.J. and Sloane, N.J.A. [1]. *The Theory of Error Correcting Codes*, North-Holland, Amsterdam, 1978.

Mahler, K. [1]. The theorem of Minkowski-Hlawka, *Duke Math. J.* **13** (1946), 611–621.

Malyšev, A.V. [1]. On the Minkowski-Hlawka theorem concerning a star body, *Uspehi Mat. Nauk* **7** (1952), 168–171.

Martinet, J. [1]. *Les Réseaux Parfaits des Espaces Euclidiens*, Masson, Paris, 1996.

Milman, V.D. [1]. Almost Euclidean quotient spaces of subspaces of finite dimensional normed spaces, *Proc. Amer. Math. Soc.* **94** (1985), 445–449.

Milnor, J. [1]. Hilbert's problem 18: On crystallographic groups, fundamental domains and on sphere packings, *Proc. Symp. Pure Math. AMS* **28** (1976), 491–506.

Minkowski, H. [1]. Sur la réduction des formes quadratiques positives quaternaires, *C. R. Acad. Sci. Paris* **96** (1883), 1205–1210. [2]. Über positive quadratische Formen, *J. reine angew. Math.* **99** (1886), 1–9. [3]. Zur Theorie der positiven quadratischen Formen, *J. reine angew. Math.* **101** (1887), 196–202. [4]. Extrait d'une lettre adressée à M. Hermite, *Bull. Sci. Math.* **17** (2) (1893), 24–29. [5]. Dichteste gitterförmige Lagerung kongruenter Körper, *Nachr. Ges. Wiss. Göttingen* (1904), 311–355. [6]. Diskontinuitätsbereich für arithmetische Äquivalenz, *J. reine angew. Math.* **129** (1905), 220–274. [7]. *Geometrie der Zahlen*, Chelsea, New York, 1953.

Mordell, L.J. [1]. Observation on the minimum of a positive quadratic form in eight variables, *J. London Math. Soc.* **19** (1944), 3–6.

Muder, D.J. [1]. Putting the best face on a Voronoi polyhedron, *Proc. London Math. Soc.* **56** (1988), 329–348. [2]. A new bound on the local density of sphere packings, *Discrete & Comput. Geom.* **10** (1993), 351–375.

Nemhauser, G.L. and Wolsey, L.A. [1]. *Integer and Combinatorial Optimization*, John Wiley & Sons, New York, 1988.

Novikova, N.V. [1]. Korkine-Zolotarev reduction domains for positive quadratic forms in $n \leq 8$ variables and reduction algorithms for these domains, *Soviet Math. Dokl.* **27** (1983), 557–560.

Odlyzko, A.M. and Sloane, N.J.A. [1]. New bounds on the unit spheres that can touch a unit sphere in n-dimensions, *J. Combinat. Theory (A)* **26** (1979), 210–214.

Ollerenshaw, K. [1]. The critical lattices of a sphere, *J. London Math. Soc.* **23** (1948), 297–299. [2]. Critical lattices of a four-dimensional hypersphere, *J. London Math. Soc.* **24** (1949), 190–200; **26** (1951), 316–318.

Ozhigova, E.P. [1]. Problems of number theory, *Mathematics of the 19th Century*, Birkhäuser, Basel, 1992, 137–210.

Pach, J. and Agarwal, P.K. [1]. *Combinatorial Geometry*, John Wiley & Sons, New York, 1995.

Rankin, R.A. [1]. On the closest packing of spheres in n dimensions, *Ann. Math.* **48** (1947), 1062–1081. [2]. On the sums of powers of linear forms I, *Ann. Math.* **50** (1949), 691–698. [3]. On sums of powers of linear forms II, *Ann. Math.* **50** (1949), 699–704. [4]. On sums of powers of linear forms III, *Proc. Kon. Ned. Akad. Wet.* **51** (1948), 846–853. [5]. On positive definite quadratic forms, *J. London Math. Soc.* **28** (1953), 309–314. [6]. The closest packing of spherical caps in n dimensions, *Proc. Glasgow Math. Assoc.* **2** (1955), 139–144. [7]. On the minimal points of positive definite quadratic forms, *Mathematika* **3** (1956), 15–24. [8]. On the minimal points of perfect quadratic forms, *Math. Z.* **84** (1964), 228–232.

Remak, R. [1]. Über die Minkowskische Reduktion der definiten quadratischen Formen, *Compositio Math.* **5** (1938), 368–391.

Rogers, C.A. [1]. A note on a theorem of Blichfeldt, *Proc. Kon. Ned. Akad. Wet.* **49** (1946), 589–594. [2]. Existence theorems in the geometry of numbers, *Ann. Math.* **48** (1947), 994–1002. [3]. A note on coverings and packings, *J. London Math. Soc.* **25** (1950), 327–331. [4]. The closest packing of convex two-dimensional domains, *Acta Math.* **86** (1951), 309–321; **104** (1960), 305–306. [5]. On a theorem of Siegel and Hlawka, *Ann. Math.* **53** (1951), 531–540. [6]. The Minkowski-Hlawka theorem, *Mathematika* **1** (1954), 111–124. [7]. Mean values over the space of lattices, *Acta Math.* **94** (1955), 249–287. [8]. A note on coverings, *Mathematika* **4** (1957), 1–6. [9]. Lattice coverings of space: the Minkowski-Hlawka theorem, *Proc. London Math. Soc.* **8** (1958), 447–465. [10]. The packing of equal spheres, *Proc. London Math. Soc.* **8** (1958), 609–620. [11]. Lattice coverings of space, *Mathematika* **6** (1959), 33–39. [12]. An asymptotic expansion for certain Schläfli functions, *J. London Math. Soc.* **36** (1961), 78–80. [13]. Covering a space with spheres, *Mathematika* **10** (1963), 157–164. [14]. *Packing and Covering*, Cambridge University Press, Cambridge, 1964.

Rogers, C.A. and Shephard, G.C. [1]. The difference body of a convex body, *Arch. Math.* **8** (1957), 220–233.

Rogers, C.A. and Zong, C. [1]. Covering convex bodies by translates of convex bodies, *Mathematika* **44** (1997), 215–218.

Royden, H. [1]. A history of mathematics at Stanford, *A Century of Mathematics in America* II, AMS, Providence, 1989, 237–277.

Rush, J.A. [1]. A lower bound on packing density, *Invent. Math.* **98** (1989), 499–509. [2]. A bound, and a conjecture, on the maximum lattice-packing density of a superball, *Mathematika* **40** (1993), 137–143.

Rush, J.A. and Sloane, N.J.A. [1]. An improvement to the Minkowski-Hlawka bound for packing superballs, *Mathematika* **34** (1987), 8–18.

Ryškov, S.S. [1]. On the reduction theory of positive quadratic forms, *Soviet Math. Dokl.* **12** (1971), 946–950. [2]. On Hermite, Minkowski and Venkov reduction of positive quadratic forms in n variables, *Soviet Math. Dokl.* **13** (1972), 1676–1679. [3]. The geometry of positive quadratic forms, *Proc. Internat. Congr. Math. Vencouver*, 1974, 501–506. [4]. Density of an (r, R)-system, *Math. Notes* **16** (1975), 855–858. [5]. The Hermite-Minkowski theory of reduction of positive definite quadratic forms, *J. Soviet Math.* **6** (1976), 651–671. [6]. On the problem of the determination of the perfect quadratic forms in many variables, *Proc. Steklov Inst. Math.* **142** (1976), 233–259.

Ryškov, S.S. and Baranovskii, E.P. [1]. Solution of the problem of the least dense lattice covering of five-dimensional space by equal spheres, *Soviet. Math. Dokl.* **16** (1975), 586–590. [2]. Classical methods in the theory of lattice packings, *Russian Math. Surveys* **34** (1979), 1–68.

Ryškov, S.S. and Cohn, M.J. [1]. On the theory of the structure of the Minkowski reduction region, *Proc. Steklov Inst. Math.* **152** (1982), 191–212.

Ryškov, S.S. and Horváth, J. [1]. Estimation of the radius of a cylinder that can be embedded in every lattice packing of n-dimensional unit balls, *Math. Notes* **17** (1975), 72–75.

Saaty, T.L. and Alexander, J.M. [1]. Optimization and the geometry of numbers: packing and covering, *SIAM Review* **17** (1975), 475–519.

Sanov, J.N. [1]. New proof of a theorem of Minkowski, *Izv. Akad. Nauk* **16** (1952), 101–112.

Schläfli, L. [1]. Réduction d'une intégrale multiple, qui comprend l'arc de cercle et l'aire du triangle sphérique comme cas particuliers, *Gesam. Math. Abh.* **2** (1855), 164–190.

Schmidt, W.M. [1]. Eine Verschärfung des Satzes von Minkowski-Hlawka, *Monatsh. Math.* **60** (1956), 110–113. [2]. Mittelwerte über Gitter, *Monatsh. Math.* **61** (1957), 269–276. [3]. Mittelwerte über Gitter II, *Monatsh. Math.* **62** (1958), 250–258. [4]. On the Minkowski-Hlawka theorem, *Illinois J. Math.* **7** (1963), 18–23.

Schneider, T. [1]. Über einen Hlawkaschen Satz aus der Geometrie der Zahlen, *Arch. Math.* **2** (1950), 81–86. [2]. Über einen Blichfeldtschen Satz aus der Geometrie der Zahlen, *Arch. Math.* **2** (1950), 349–353.

Schütte, K. and van der Waerden, B.L. [1]. Das Problem der dreizehn Kugel, *Math. Ann.* **125** (1953), 325–334.

Seeber, L.A. [1]. *Untersuchungen über die Eigenschaften der positiven ternären quadratischen Formen*, Freiberg, 1831.

Segre, B. and Mahler, K. [1]. On the densest packing of circles, *Amer. Math. Monthly* **51** (1944), 261–270.

Seidel, J.J. [1]. Harmonics and combinatorics, special functions, *Group Theoretical Aspects and Applications*, Reidel, Dordrecht, 1984, 287–303.

Selling, E. [1]. Über die binären und ternären quadratischen Formen, *J. reine angew. Math.* **77** (1873/74), 143–229.

Shannon, C.E. [1]. Probability of error for optimal codes in a Gaussian channel, *Bell System Tech. J.* **38** (1959), 611–656.

Shioda, T. [1]. Mordell-Weil lattices and sphere packings, *Amer. J. Math.* **113** (1991), 931–948.

Sidelnikov, V.M. [1]. On the densest packing of balls on the surface of the n-dimensional Euclidean sphere, and the number of vectors of a binary code with prescribed code distances, *Soviet Math. Dokl.* **14** (1973), 1851–1855. [2]. New estimates for the closest packing of spheres in n-dimensional Euclidean space, *Mat. Sb.* **95** (1974), 148–158.

Siegel, C.L. [1]. A mean value theorem in geometry of numbers, *Ann. Math.* **46** (1945), 340–347. [2]. *Lectures on the Geometry of Numbers*, Springer-Verlag, Berlin, 1989.

Skubenko, B.F. [1]. Dense lattice packings of spheres in Euclidean spaces of dimension $n \leq 16$, *J. Soviet Math.* **18** (1982), 958–960. [2]. A remark on the upper bound of the Hermite constant for the densest lattice packing of spheres, *J. Soviet Math.* **18** (1982), 960–961.

Sloane, N.J.A. [1]. Sphere packings constructed from BCH codes and Justesen codes, *Mathematika* **19** (1972), 183–190. [2]. Binary codes, lattices, and sphere-packings, *Proc. Sixth British Combinat. Conf.*, Academic Press, London, 1977, 117–164. [3]. Tables of sphere packings and spherical codes, *IEEE Trans. Inform. Theory* **27** (1981), 327–338. [4]. Recent bounds for codes, sphere packings and related problems obtained by linear programming and other methods, *Contemp. Math.* **9** (1982), 153–185. [5]. The packing of spheres, *Scientific American* **250** (1) (1984), 116–125.

Stacey, K.C. [1]. The enumeration of perfect septenary forms, *J. London Math. Soc.* **10** (1975), 97–104.

Štogrin, M.I. [1]. The Voronoi, Venkov and Minkowski reduction domains, *Soviet Math. Dokl.* **13** (1972), 1698–1702. [2]. Locally quasi-densest lattice packing of spheres, *Soviet Math. Dokl.* **15** (1974), 1288–1292.

Stuhler, U. [1]. Eine Bemerkung zur Reduktionstheorie quadratischer Formen, *Arch. Math.* **27** (1976), 604–610. [2]. Zur Reduktionstheorie der positiven quadratischen Formen II, *Arch. Math.* **28** (1977), 611–619.

Swinnerton-Dyer, H.P.F. [1]. Extremal lattices of convex bodies, *Proc. Cambridge Phil. Soc.* **49** (1953), 161–162.

Szegö, G. [1]. *Orthogonal Polynomials*, Amer. Math. Soc., Providence, 1975.

Talata, I. [1]. Exponential lower bound for the translative kissing numbers of d-dimensional convex bodies, *Discrete & Comput. Geom.* **19** (1998), 447–455. [2]. The translative kissing number of tetrahedra is eighteen, *Discrete & Comput. Geom.*, in press. [3]. Translational clouds of convex bodies, preprint.

Tammela, P.P.   [1].   On the reduction theory of positive quadratic forms, *Soviet Math. Dokl.* **14** (1973), 651–655. [2]. The Hermite-Minkowski domain of reduction of positive definite quadratic forms in six variables, *J. Soviet Math.* **6** (1976), 677–688. [3]. On the reduction theory of positive quadratic forms, *J. Soviet Math.* **11** (1979), 197–277. [4]. The Minkowski reduction domain for positive quadratic forms of seven variables, *J. Soviet Math.* **16** (1981), 836–857.

Temesvári, A.H., Horváth, J., and Yakovlev, N.N.   [1]. A method for finding the densest lattice k fold packing of circles, *Math. Notes* **41** (1987), 349–355.

Thompson, T.M.   [1].   *From Error-Correcting Codes Through Sphere Packings to Simple Groups*, Math. Assoc. Amer., 1983.

Thue, A.   [1].   Om nogle geometrisk taltheoretiske theoremer, *Forhdl. Skand. Naturforsk.* **14** (1892), 352–353. [2]. Über die dichteste Zusammenstellung von kongruenten Kreisen in der Ebene, *Norske Vid. Selsk. Skr.* **1** (1910), 1–9.

Tsfasman, M.A.   [1].   Algebraic curves and sphere packings, *Arithmetic, Geometry and Coding Theory*, de Gruyter, Berlin, 1996, 225–251.

Tsfasman, M.A. and Vladuts, S.G.   [1]. *Algebraic-Geometric Codes*, Kluwer, Dordrecht, 1991.

Vardy, A.   [1].   A new sphere packing in 20 dimensions, *Invent. Math.* **121** (1995), 119–133.

Venkov, B.A.   [1].   Über die Reduktion positiver quadratischer Formen, *Izv Akad. Nauk. SSSR Ser. Mat.* **4** (1940), 37–52.

Vetčinkin, N.M.   [1].   The packings of uniform n-dimensional balls that are constructed on error-correcting codes, *Ivanov. Gos. Univ. Učen. Zap.* **89** (1974), 87–91. [2]. Uniqueness of the classes of positive quadratic forms on which the values of the Hermite constants are attained for $6 \le n \le 8$, *Proc. Steklov Inst. Math.* **152** (1982), 37–95.

Voronoi, G.F.   [1].   Nouvelles applications des parammètres continus à la théorie des formes quadratiques. Premier Mémoire. Sur quelques propriétés des formes quadratiques positives parfaites, *J. reine angew. Math.* **133** (1908), 97–178. [2]. Deuxième Mémoire. Recherches sur les paralléloèdres primitifs, *J. reine angew. Math.* **134** (1908), 198–287. [3]. Domaines de formes quadratiques corresponant aux différent types des paralléloèdres primitifs, *J. reine angew. Math.* **135** (1909), 67–181.

van der Waerden, B.L.   [1].   Die Reduktionstheorie der positiven quadratischen Formen, *Acta Math.* **96** (1956), 265–309.

van der Waerden, B.L. and Gross, H.   [1]. *Studien zur Theorie der quadratischen Formen*, Birkhäuser, Basel, 1968.

Watson, G.L. [1]. The covering of space by spheres, *Rend. Circ. Mat. Palermo* **5** (1956), 93–100. [2]. On the minimum of a positive quadratic form in n (≤ 8) variables, verification of Blichfeldt's calculations, *Proc. Cambridge Phil. Soc.* **62** (1966), 719. [3]. On the minimum points of a positive quadratic form, *Mathematika* **18** (1970), 60–70. [4]. The number of minimum points of positive quadratic forms, *Dissertationes Math.* **84** (1971), 1–43.

Weil, A. [1]. Sur quelques résultats de Siegel, *Summa Brasiliensis Math.* **1** (1946), 21–39.

Wills, J.M. [1]. Research problem 35, *Period. Math. Hungar.* **14** (1983), 312–314. [2]. On the density of finite packings, *Acta Math. Acad. Sci. Hungar.* **46** (1985), 205–210. [3]. An ellipsoid packing in E^3 of unexpected high density, *Mathematika* **38** (1991), 318–320. [4]. Finite sphere packings and sphere coverings, *Rend. Semin. Mat. Messina* **2** (1993), 91–97. [5]. Finite sphere packings and the methods of Blichfeldt and Rankin, *Acta Math. Acad. Sci. Hungar.* **74** (1997), 337–342.

Woods, A.C. [1]. The densest double lattice packing of four spheres, *Mathematika* **12** (1965), 138–142. [2]. Lattice coverings of five space by spheres, *Mathematika* **12** (1965), 143–150. [3]. Covering six space with spheres, *J. Number Theory* **4** (1972), 157–180.

Wyner, J.M. [1]. Capabilities of bounded discrepancy decoding, *Bell System Tech. J.* **44** (1965), 1061–1122. [2]. Random packings and coverings of the unit n-sphere, *AT&T Tech. J.* **46** (1967), 2111–2118.

Yakovlev, N.N. [1]. On the densest lattice 8-packing in the plane, *Moscow Univ. Math. Bull.* **38** (1983), 7–16.

Yang, L.J. [1]. Multiple lattice packings and coverings of spheres, *Monatsh. Math.* **89** (1980), 69–76.

Yudin, V.A. [1]. Sphere-packing in Euclidean space and extremal problems for trigonometrical polynomials, *Discr. Appl. Math.* **1** (1991), 69–72.

Zong, C. [1]. On a conjecture of Croft, Falconer and Guy on finite packings, *Arch. Math.* **64** (1995), 269–272. [2]. Some remarks concerning kissing numbers, blocking numbers and covering numbers, *Period. Math. Hungar.* **30** (1995), 233–238. [3]. The kissing numbers of tetrahedra, *Discrete & Comput. Geom.* **15** (1996), 239–252. [4]. *Strange Phenomena in Convex and Discrete Geometry*, Springer-Verlag, New York, 1996. [5]. The translative kissing numbers of the Cartesian product of two convex bodies, one of which is two-dimensional, *Geom. Dedicata* **65** (1997), 135–145. [6]. A problem of blocking light rays, *Geom. Dedicata* **67** (1997), 117–128. [7]. On the kissing numbers of convex bodies–A brief survey, *Bull. London Math. Soc.* **30** (1998), 1–10. [8]. A note on Hornich's problem, *Arch. Math.* **72** (1999), 127–131.

Index

Universitext   *(continued)*